D1153973

The Car Design Yearbook 1

Mercedes-Benz

Stephen Newbury

The Car Design Yearbook

the definitive guide to new concept and production cars worldwide

MERRELL

BB X 252

MERRELL

First published 2002 by
Merrell Publishers Limited
42 Southwark Street
London SE1 1UN

Telephone + 44 (0)20 7403 2047
E-mail mail@merrellpublishers.com

Publisher Hugh Merrell
Editorial Director Julian Honer
Managing Editor Anthea Snow
Art Director Matthew Hervey
Production Manager Kate Ward
Editorial and Production Assistant Emily Sanders
Sales and Marketing Manager Emilie Nangle
Sales and Marketing Executive Eddy Obermüller

    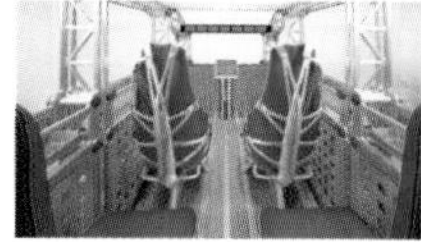   

Distributed in the USA by Rizzoli International Publications, Inc. through St Martin's Press, 175 Fifth Avenue, New York, NY 10010

A catalog record for this book is available from the Library of Congress

ISBN 1 85894 190 3

Consultant editors: Giles Chapman, Tony Lewin, Julian Rendell, Karl Ludvigsen
Edited by Iain Ross and Laura Hicks
Designed by Kate Ward
Printed and bound in Spain

Frontispiece: Mercedes-Benz CLK
pp. 4–5: MCC Smart Crossblade
pp. 8–9: Mercedes-Benz Vision GST

Contents

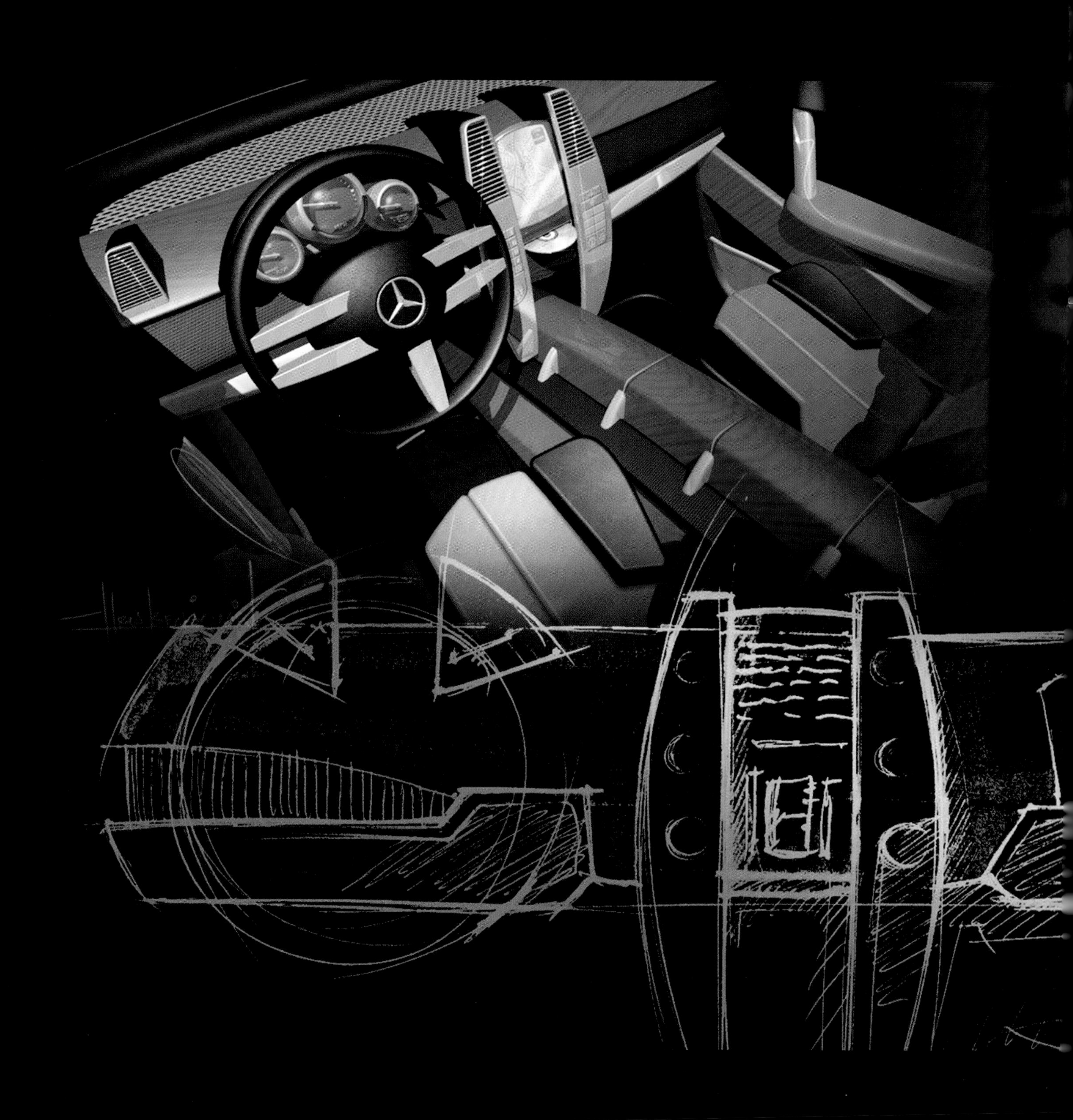

nitial design theme
SRS

Trends, Highlights, Predictions

This, the first *Car Design Yearbook*, features every significant new concept and production car unveiled at auto shows across the world from April 2001 through March 2002. Only cars that are completely new, or that have an overwhelming element of new design, are included. You will not find mid-cycle or superficial facelifts, launched merely to stem a downturn in sales or extend a model's life. Nor have we included derivatives of established new models, such as wagons or convertibles, often introduced years after the mainstream car went on sale.

The Car Design Yearbook focusses only on new and future car design. This makes it a unique annual event in both design and automotive publishing—the first time ever that such a comprehensive review and critique of new vehicles has been collated in one book. To give you an idea of the scale of the task, the past twelve months have seen the unveiling of 127 totally new concept and production cars, and all of them are covered here. *The Car Design Yearbook* will become the definitive guide to concept and production cars worldwide.

The reviews are laid out alphabetically in order of brand, then model. Accompanying each model is an explanation of the design and its significance to the make, together with a specification table listing its key technical features. Some concepts remain merely clay models, so the information available is limited, as the car clearly will not run. Each year the book will also review three influential designers of the past year. Overviews of their careers and discussions of several of their most important works will be accompanied by pictures of their latest and past achievements. This year the profiles are of Ian Callum, Giorgetto Giugiaro, and Freeman Thomas.

The driving force behind the evolution of car design is explained partly by examining the social, economic, and political factors that directly affect the way we live, and partly by investigating the technological innovations that present new opportunities for designers. In Europe, escalating fuel prices and intensifying road congestion have meant extra demand for small cars. These models are generally well engineered and quality-built, giving high levels of refinement and safety. Thanks to hatchbacks and folding

Opposite from top

Volkswagen's new Polo is a typical example of the modern, small family car: a spacious hatchback contained within compact overall proportions, yet exuding a Germanic, well-built aura

If car design can be said to typify an era, in the way the original Mini evokes the 1960s, and the first Renault Espace the 1980s, then maybe the irreverent MCC Smart—in this case, the limited-edition Crossblade—sums up the first decade of the twenty-first century

The new Range Rover, as far removed from an agricultural vehicle as it is possible to get, panders to the market's addiction to materialism, while offering excellent visibility and security

rear seats, they offer large load-carrying capacity, despite compact overall dimensions.

However, in Western cities, car ownership is almost at saturation point. Road tolls, and the banning of cars from city centers, are likely to force people toward public transportation, and governments are working busily on attractive alternatives to the car by developing integrated transportation systems. The introduction in inner-city areas of zero-emission vehicles, to reduce pollution, may also be given priority. Still, the comfort, freedom, flexibility, security, and privacy the car offers make it very difficult to replace.

In the US, with the new Bush administration, it appears unlikely that a strong green agenda will prevail. We can expect low fuel prices to continue, and large-displacement vehicles to keep selling well. And the massive potential consumer markets in China and India have barely been tapped.

Car design influences the lives of millions of people throughout the world. Whether the car serves as merely a practical means of transport, or as an extension of one's personality, its design and brand will always attract comment. The design of cars has long been symbolic of our times. Alongside other branches of the applied arts, such as architecture and fashion, cars are lasting reminders of the characters of different decades. Those who recall the fin-tailed Cadillac of the 1950s, the Mini of the 1960s, or the Ford Capri of the 1970s, will recognize how significant cars are as mirrors of contemporary fashion trends. Perhaps we will look back on 2002 with wistful memories of the Smart City Coupé.

Car manufacturers seem to think we spend our weekends on activity pursuits. There has certainly been an explosion of sports utility vehicles (SUVs) on the market. This addiction to off-road vehicles is set to continue. Our obsession with materialism and status—"the bigger the car, the richer we feel" syndrome—means there is a lucrative market for these models while the economy remains resilient. Most of us do not need the off-road capability of a Range Rover while taking the kids to school, but we enjoy the good visibility and security of its commanding driving position.

Car designs are not always executed by the manufacturers themselves. They concentrate on creating the high-volume, mainstream products that provide their "bread and butter," while such

The Italdesign Brera concept, produced for Alfa Romeo, is a fine example of the close working relationship that exists between car manufacturers and outside design consultancies

specialist car-design consultancies as TWR, Italdesign, and Pininfarina apply their nimble, sophisticated talents to bringing new products to market appropriately and rapidly. It may seem strange that prestigious brands will entrust "outsiders" with the look of a future model, but there are several very good reasons for this. A manufacturer's in-house design team may simply be too busy with current projects, and have to outsource one to keep every project on schedule. A car maker may want a specific style injected into its products; Pininfarina, for instance, was commissioned to design the Peugeot 306 convertible not only because its experience in designing convertible roofs made it an ideal trustee for this complex work, but also for the prestige of Peugeot's being associated with the consultancy, owing to the latter's historic involvement with many exotic cars—especially Ferraris. Unsurprisingly, every 306 convertible has a discreet Pininfarina emblem fitted just in front of the rear wheel arch.

If a manufacturer wants to embark on a totally new style direction, a consultancy can prove valuable because it will have worked for many car companies and be used to thinking laterally and coming up with imaginative ideas. Ford, for instance, teamed up with Pininfarina to create the Start concept car shown at Frankfurt in 2001, and Alfa Romeo often uses Italdesign, which designed the Brera, one of the most highly acclaimed concepts at the 2002 Geneva show. The work of consultancies is often uncredited, because confidentiality contracts drawn up by manufacturers allow them alone to claim responsibility for the design once it is in the public domain.

A concept car can cost in the region of $4,500,000 to create. This may seem an incredible extravagance if the model fails to make it as a production car. Concepts are, however, vital to the development and success of a brand: they test the press and public reaction to a new model or styling theme. When you realize that Ford or GM might invest around $1,500,000,000 in the development of a new model, you can see how important it is for them to be sure that the market wants their product. Usually it takes anything from one to four years to see a concept model reach production, depending on the stage of the development process at which the concept is first shown.

New life can be breathed into tired car brands through the medium of exciting design, as Chrysler has proved with its Prowler, PT Cruiser, and now, shown here, its Crossfire coupe

It is absolutely right to wonder why some cars from different brands look so similar. The commercial benefit of sharing components is enormous. For example, the VW group's Volkswagen, Seat, and Skoda brands have made huge profits from component sharing; they all offer cars using identical "platforms"—the fully functioning mechanical underpinnings. Within the industry, however, some concern has been expressed as to whether Volkswagen can truly differentiate between its brands while they share such intrinsic elements. So far it has managed this extremely well, but all its good work can be undone very quickly if the automotive press decrees, for example, that a Volkswagen is nothing more than a Skoda with a VW emblem—and a 20% price premium.

The car-design industry can also be incestuous. Design leaders command high salaries. They are often poached between brands to counter sagging sales with fresh thinking. The movements of design leaders during the last year demonstrate this constant cross-fertilization: Henrik Fisker is now creative director of Ford's London Design Enterprise, and new design director for Aston Martin, taking over from Ian Callum; before joining Ford, this Danish designer was president and chief executive officer of Designworks/USA, a BMW subsidiary, so it will be intriguing to see what he will do with the very British Aston Martin. Meanwhile, California-based Designworks/USA has appointed Adrian van Hooydonk as president. He will report directly to Chris Bangle, director of design at BMW. Moray Callum is the new head of global design for Mazda, with a mission to develop further the Ford-controlled Japanese manufacturer's design philosophy: "Emotion in Motion."

Fiat Auto has set up an independent design studio outside Europe, in Brazil. Peter Fassbender, who had been chief designer of exteriors at Fiat since 1996, has been assigned to run the new studio. Walter de Silva now has responsibility for design within the Audi group, which also encompasses the Lamborghini and Seat brands. Peter Schreyer, previously responsible for Audi design, is now head of design for the Volkswagen brand. Freeman Thomas has been made head of the Pacifica Advance Product Design Center in Carlsbad, California, for Chrysler. The American arm of DaimlerChrysler has

had big successes recently with the Crossfire and PT Cruiser, so Thomas has a firm base to build on. Andreas Zapatinas, the Greek designer, is the new head of advanced design for Subaru, which makes him the second European to run the design center of a Japanese car maker—Olivier Boulay leads Mitsubishi's design centers worldwide.

Because lead designers move around the industry so rapidly, no one brand can really get too far ahead of its competition. In fact, because there has been so much consolidation, companies such as Ford, with its Premier Automotive Group (encompassing Land Rover, Volvo, Jaguar, and Aston Martin), have a broad, global view of design trends in any case.

It is interesting to look at the progression of design within a brand. Jaguar's R-Coupé, for example, claims to signpost the design direction for all future Jaguars. Its barreled shape is reminiscent of Jaguar design of the 1960s. As in those now far-off days, its complex, emotional surfaces set it apart. Yet Jaguar, under pressure from its owner, must use common components from higher-volume Ford models to reduce costs. It is a skilled balancing act to maintain its clear identity as a luxury British car maker in the face of such bean-counter-led pressure. The controversial new Jaguar X-type shares its platform with the Ford Mondeo, making it the first front-wheel-drive Jaguar, sailing close to the wind in credibility terms.

The trend for "crossover" vehicles continues, some combining the utility of larger vehicles with the ruggedness of off-roaders and the performance of sports cars. A good current example is the BMW X5. It has exceeded market-analysis predictions, and is a strong seller in the US and Europe. Many new concepts for "crossover" vehicles were shown last year, including Mercedes-Benz's GST, Saab's 9X, Infiniti's FX45, and Volkswagen's Magellan.

The desire for more spacious, well-lit, and relaxing interiors has led to more new models being slightly wider and taller than those they replaced. This expansion of each model also allows manufacturers to introduce new, smaller models at the bottom of their ranges. This pushes existing cars further upmarket—where, of course, business is more lucrative. Typical recent European examples of this are the Ford Ka and Volkswagen Lupo.

Increased attention to the interior has been a hallmark of many concept cars over the past year. Designers of some concepts have chosen to ignore the practicalities of driving comfort or safety to explore the versatility of an interior. Seats that fold into beds, or rotate, to create a living-room environment have been tried. Renault is a leader in the field of interior design. It combines harmonious colors, materials and elegantly formed surfaces with increased glass in the roof to create a relaxing atmosphere. Audi executes interior design with equal excellence, although the result is usually darker: solid, classic-looking, and exuding genuine quality.

Highlights of the past year have come from across the globe. The Renault Talisman concept displayed the new corporate face of Renault, with huge "gull-wing" doors; it is an outrageous car with unique proportions. The Seat Tango was a gorgeous little concept that oozed Mediterranean flavor, a car that would instantly draw people into Seat showrooms. It will be interesting to see if Volkswagen gives the green light for a production version. The new Maserati Coupé and Spyder are beautiful cars, and instantly recognizable as Italian designs; they are sure to be big successes, and meet their sales targets.

Daihatsu showed a neat little sports car called the Copen, with a folding hard-top. It would be surprising not to see a production version. The Toyota pod concept, with its unique ability to show the driver's emotions, was jointly developed with Sony. It is sure to make designers reconsider the relationship between driver and car. The Chrysler Crossfire production version, although differing slightly from the original concept, retains its unique identity, and should be a big hit for Chrysler. Also, from DaimlerChrysler, the Mercedes-Benz GST concept looks fantastic—it has amazing presence. This car would be a much better competitor to the BMW X5 than Mercedes-Benz's current M-Class. We will surely see a production version.

At Geneva in 2002, Maybach made its comeback. This enormous automobile demonstrated the heights of possible luxury. It is unlikely to be driven by anyone other than a chauffeur. This is a real sign that people are still prepared to invest huge amounts of money in cars. Also at Geneva was the new

Opposite
The Jaguar R-Coupé concept seeks to define a new direction in design for the British brand by raiding the archives, in this case for the "barreled" look of great Jaguars of the 1960s, such as the MkII sedan

BMW CS1 concept, a small convertible given a new style by Chris Bangle. This concept was well received, and a showroom edition will follow.

The current trend for large sports utility vehicles (SUVs), together with designers wanting to fit the largest possible wheel/tire combinations to give a car a strong persona, has led to advances in wheel and tire design. It is now common to see 19 ins wheels fitted on SUVs and large sedans. Technological advances in lighting, particularly with LEDs at the rear, and Xenon lamps at the front, give designers the chance to use less surface area for lamps, because their light intensity is higher, and their beams can be adjusted automatically.

Chassis systems have been a strong focus of development for Mercedes-Benz in the past year. The F400 Carving has a suspension that tilts its wheels and specially developed tires into the corner, like a cross between a motorcycle and a car. This boosts cornering forces to give enhanced active safety. Mercedes is also now fitting its advanced active-braking system (ABS) to its lesser models as well as its top-line offerings: ABS immediately pressurizes the brake system when an emergency situation is suspected, cutting braking distances. Folding metal roof systems, as fitted to the Mercedes-Benz SL and SLK, and the Peugeot 206, have achieved widespread market acceptance. This has paved the way for new roof-system concepts that transform the upper architecture of a car for other useful functions, such as carrying large loads. Expect to see more of these in the future.

In Europe, tough new legislation is in place to reduce vehicle emissions and fuel consumption, but in America the CAFE (Corporate Average Fuel Economy) rules are still set sufficiently low, and with enough get-out clauses, to ensure that the big three US manufacturers—General Motors, DaimlerChrysler, and Ford—can still produce millions of gas-guzzling trucks each year. Reducing aerodynamic drag is one method of improving fuel consumption, and a way of doing this is to create a streamlined shape that waists gradually at the rear—as in the Daihatsu UFE. Two other concept cars, which will be launched officially at auto shows late in 2002, share a similar design objective, employing sleek, aerodynamic shapes, and lightweight carbon-fiber structures. They are, however, at completely opposite ends of the

performance and fuel-economy spectrum. The Volkswagen L1 has a 300cc diesel engine, and an extremely aerodynamic package, to propel it 100 km (62 miles) using only a single liter of diesel fuel. The Ferrari FX45, on the other hand, is a high-performance car with an extremely aerodynamic shape, borrowing this aspect of its design from Ferrari's Formula 1 cars. If these cars were put into production, it is likely that their styling and packaging would be compromised. Thankfully, attractive design still takes priority.

In the long term, legislation will affect the cars we drive. Increased emissions-related taxes, and spiraling fuel prices, will make us all reconsider whether we need big cars, when passenger occupancy rates have fallen to below two people per car in Europe and the US. Such concepts as the 3PV (3-Person Vehicle), designed by TWR's Neil Simpson, address this problem, and are a welcome reminder that small cars need not be conventionally styled hatchbacks.

In the short term, though, we can expect a quarter of the concepts analyzed in this year's *Car Design Yearbook* to make it as production cars. The others will remain pure design exercises, and perhaps lend their better styling features to future models.

Opposite from top
Maserati plays on its traditional design strengths of sensual, curvaceous forms for its new coupe, no doubt matching buyer expectations

BMW demonstrates its intention of entering the small-car sector—the domain of the Fiat Punto—with its skillfully executed CS1 concept

The dramatic Mercedes-Benz F400 Carving is as much a rolling technology testbed as a crowd-pleasing auto-show centerpiece

Left from top
The Ferrari FX45, likely to be dubbed "F60" when it goes on sale, uses Formula 1-style aerodynamics to obtain ultra-high speeds with stability

The 3PV, created by TWR, tackles the issue of low vehicle occupancy by offering just three seats in a compact car

Volkswagen's incredible L1 economy concept car mixes light weight and aerodynamic efficiency to travel more than 200 miles on a gallon of diesel fuel

Acura RD-X
Audi Avantissimo
Bertone Novanta
BMW 7 Series
BMW CS1
Cadillac Cien
Cadillac CTS
Cadillac XLR
Chevrolet Bel Air
Chevrolet SSR
Chrysler Crossfire
Chrysler Pacifica
Citroën C3
Citroën C8
Citroën C-Crosser
Daewoo Kalos
Daihatsu Copen
Daihatsu FF Ultra Space
Daihatsu Muse
Daihatsu U4B
Daihatsu UFE
Dodge M80
Dodge Razor
Fiat Ulysse
Fioravanti Yak
Ford F-350 Tonka
Ford Fiesta
Ford Fusion
Ford GT40
GM AUTOnomy
Honda Bulldog
Honda CR-V
Honda Dualnote
Honda Pilot
Honda Unibox
Hummer H2
Hyundai Clix
Hyundai Getz
Hyundai Tiburon Coupé
Infiniti FX45
Infiniti G35
Irmscher Inspiro
Isuzu GBX
Isuzu Zen
Italdesign Brera
Jaguar R-Coupé
Jeep Compass
Kia Sorento
Koenigsegg CC 8S
Lamborghini Murciélago
Lancia Phedra
Lexus GX470
Lexus Movie
Lincoln Continental
Lincoln MK 9
Lincoln Navigator
Maserati Coupé
Maserati Spyder
Matra m72
Maybach
Mazda6/Atenza
Mazda MX Sport Runabout
Mazda RX-8
Mazda Secret Hideout
MCC Smart Crossblade
MCC Smart Tridion4
Mercedes-Benz CLK
Mercedes-Benz E-Class
Mercedes-Benz F400 Carving
Mercedes-Benz SL
Mercedes-Benz Vaneo
Mercedes-Benz Vision GST
MG X80
Mitsubishi CZ2
Mitsubishi Pajero Evolution 2+2
Mitsubishi Space Liner
Mitsubishi SUP
Nissan Crossbow
Nissan GT-R
Nissan Ideo
Nissan Kino
Nissan mm.e
Nissan Moco
Nissan Nails
Nissan Quest
Nissan Yanya
Nissan Z
Opel Concept M
Opel Frogster
Opel Signum2
Opel/Vauxhall Vectra
Peugeot 807
Peugeot RC
Pininfarina Ford Start
Pontiac Solstice
Pontiac Vibe
Range Rover
Renault Espace
Renault Talisman
Rinspeed Presto
Rover TCV
Saab 9X
Saab 9-3X
Seat Tango
Skoda Superb
Skoda Tudor
Subaru Baja
Suzuki Aerio
Suzuki SX
Th!nk City
Toyota ccX
Toyota Corolla
Toyota DMT
Toyota ES³
Toyota FXS
Toyota Ist
Toyota pod
Toyota RSC
Toyota UUV
Toyota Voltz
Toyota WiLL VC
Venturi Fétish
Volkswagen Magellan
Volkswagen Phaeton
Volkswagen Polo
Volkswagen W12 Coupé
Volvo XC90

A–Z of New Models

Acura RD-X

Design	Ricky Hsu
Engine	2.4 in-line 4
Power	186 kW (250 bhp)
Gearbox	6-speed clutchless manual
Installation	Front-engine/4-wheel drive
Brakes front/rear	Discs/discs
Front tires	235/60R18
Rear tires	235/60R18
Length	4265 mm (167.9 ins)
Width	1900 mm (74.8 ins)
Height	1561 mm (61.5 ins)
Wheelbase	2573 mm (101.3 ins)

This concept car, the RD-X, is clearly only a styling exercise, a fantasy idea for a sports utility vehicle intended to flavor the ruggedness of a traditional SUV with sports-car performance. In doing this, Acura—the brand Honda created in 1986—has tried to meet the lifestyle needs of that desirable stratum of consumers: active urbanites.

The RD-X is a dramatic design. Its dominant features are a high belt line, and huge wheel arches. The glass roof gives maximum visibility, and the rear roof panel opens so that tall items can be transported. The loadspace versatility is increased by side doors that both open wide, and, with no central pillar to get in the way, give fantastic access to the interior.

At the rear, the RD-X features two powered doors, uniquely designed, that slide open to the sides, giving wide access for bicycles or snowboards. Inside, the RD-X has a futuristic, driver-oriented cockpit as part of a large, easily transformable cargo area.

RD-X technology includes a head-up display that projects information such as speed and fuel level on to the windshield, so that the driver can keep his eyes focussed on the road. Instead of side-view mirrors, the car employs two rear-facing cameras mounted on the front fenders. They transmit a rearward view of the traffic to displays on the steering column.

There are some questionable design omissions, such as the lack of seat belts, and a solution as to how the glass would drop into the door frame. Otherwise, this is a promising design.

Audi Avantissimo

Design	Peter Schreyer and Lutz Sauvant
Engine	4.2 V8 twin-turbo
Power	321 kW (430 bhp)
Torque	600 Nm (443 lb ft)
Gearbox	6-speed automatic/6-speed manual
Installation	Front-engine/all-wheel drive
Front suspension	4-link layout with air springs
Rear suspension	Trapezoidal link layout with air springs
Brakes front/rear	Discs/discs
Front tires	22 ins PAX run-flat system
Rear tires	22 ins PAX run-flat system
Length	5060 mm (199.2 ins)
Width	1910 mm (75.2 ins)

Audi's determination to dominate in the luxury-car stakes—the "D" sector in marketing terms, typified by the supremely capable Lexus LS430—knows no bounds. The aluminum-bodied A8 has come close to breaking the stranglehold Mercedes-Benz exerts on the sector. With the amazing Avantissimo, Audi is previewing a brand-new approach to "top of the range cars." It's a 5 m long (16.4 ft), luxury, high-performance car set to test the perception of a wagon-style car at the upper end of the market.

A light aluminum spaceframe (in Audi-speak, an ASF) is used for the structure, which is based on the A8 sedan introduced in 1994. This helps to minimize the body's additional weight, and so maximize dynamic performance. A series of glass roof panels fitted with photovoltaic elements allows controlled levels of natural light to enter the cabin. These are used as an active design element, increasing passenger comfort, and the perception of spaciousness. They also capture and store solar energy, which is then used to power the car's cooling system in hot weather.

Each passenger has an individual, leather-upholstered, wood-backed seat. The seats' width, and sense of relaxation, evoke the cossetting comfort of business-class air travel. The Multi Media Interface (MMI) is Audi's new ergonomic control system, bringing together all driver information and communication features in a single unit, and displaying data on a monitor in the driver's direct field of vision. Potential new technology is exploited with an ambitious driver ID system that incorporates a fingerprint scanner. This automatically selects the seat position, air-conditioning setting, and MMI configuration each driver has personally chosen in the past.

The Avantissimo is contemporary in style, and builds on Audi's reputation for capable station wagons, currently culminating in the A6 Avant. The Avantissimo also challenges other brands producing high-performance wagons—such as Volvo, and even BMW—to provide anything quite as luxurious and spacious. However, it remains to be seen whether Audi, as one of Volkswagen's luxury brands, is brave enough to put the Avantissimo into production.

Avantissimo

Avantissimo

Bertone Novanta

The Novanta concept, from venerable Italian coachbuilder Bertone, explores the future relationship between driver and car. Its design mission is to demonstrate the rapidity with which the driver's personal, detailed preferences can be accommodated by the car's systems and ergonomics.

At the heart of this new interface is a Nokia Communicator cell phone. Functioning as databank, on-line communication resource, and "smart access" control key, the Nokia Communicator has its own docking point in the Novanta. When the driver approaches the car, it automatically recognizes his phone, unlocks the driver's door, and activates the personalized settings for functions such as the audio and air-conditioning.

Housed in the dash is a biometric scanning device that recognizes a fingerprint. Once stored, the image becomes a unique password, providing a secure "ignition key" that also recognizes, and sets, an individual's comfort preferences.

The body surfaces originate from concave and convex volumes underlined by tense, geometric lines. Its smooth, vertical flanks are defined by a high belt line and sloping roof line. At the back, the roof forms almost a right angle with the large, vertical rear window, dropping sharply to the Novanta's rear. The exterior architecture moves the visual center of gravity toward the rear axle. A sculptured shoulder line runs the length of the car, appearing to propel it forward.

A particularly striking feature of the Novanta's interior is a luminous strip that runs the entire mid-line of the passenger compartment. Dropping to floor level immediately in front of both front and rear seats, it displays information from the comprehensive driving and navigation aids, plus audio, and the climate control.

Bertone celebrates its ninetieth anniversary as an Italian design institution with the Novanta, a concept car that combines advanced design with pioneering technology. Saab has offered its logo for this model, as proof of the two companies' strong working relationship. However, it is only the latest in a very long line of attention-grabbing show-stoppers, few of which have been seen in the hands of customers.

Engine	3.0 V6
Power	147 kW (200 bhp) @ 5000 rpm
Torque	310 Nm (229 lb ft) @ 2200 rpm
Gearbox	Automatic
Brakes front/rear	Electro-mechanical
Front tires	245/40R20
Rear tires	245/40R20
Length	4450 mm (175.2 ins)
Width	1800 mm (70.9 ins)
Height	1450 mm (57.1 ins)

SAAB

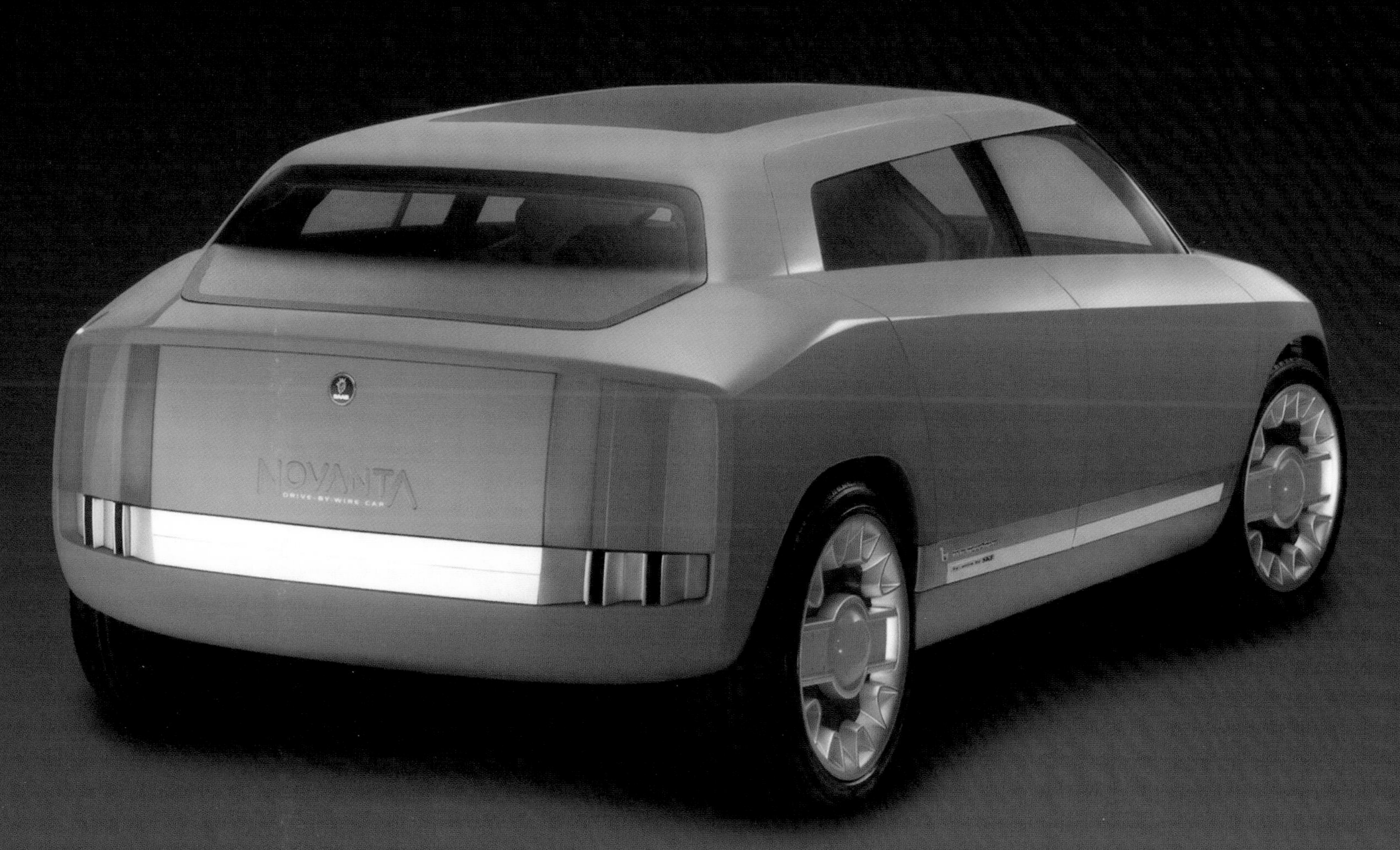
NOVANTA
DRIVE-BY-WIRE CAR

BMW 7 Series

Design	Chris Bangle
Engine	4.4 V8 (3.5 V8 and 3.0 in-line 6 diesel also offered)
Power	245 kW (333 bhp) @ 6100 rpm
Torque	450 Nm (332 lb ft) @ 3600 rpm
Gearbox	6-speed automatic with Steptronic function
Installation	Front-engine/rear-wheel drive
Front suspension	Double-joint, thrust-rod spring strut axle
Rear suspension	Multi-beam independent
Brakes front/rear	Discs/discs
Front tires	245/55R17
Rear tires	245/55R17
Length	5029 mm (198 ins)
Width	1902 mm (74.9 ins)
Height	1492 mm (58.7 ins)
Wheelbase	2990 mm (117.7 ins)
Track front/rear	1586/1590 mm (62.4/62.6 ins)
Curb weight	1945 kg (4289 lb)
0–100 km/h (62 mph)	6.3 sec.
Top speed	250 km/h (155 mph)
Fuel consumption	10.9 ltr/100 km (21.5 mpg)
CO_2 emissions	263 g/km

BMW's 7 Series, designed under the direction of Chris Bangle, is a new car in every respect. The key elements of its design are sporting performance, dynamic style, a luxurious drive, and a strong presence.

Superficially, its overall shape resembles the previous model. Close up, however, it is a notably different car. At the front, machined-aluminum headlamps and horizontal turn signals sit behind a glass cover. The trunk has been raised for improved luggage capacity and aerodynamic stability. The side view retains the traditional wedge shape, but features a rather disjointed area at the base of the rear window.

BMW's three core sedan ranges, the 3, 5, and 7 Series, are like Russian dolls: three sizes of essentially the same design that keep customers loyal no matter how much their disposable wealth balloons. That means that, whether or not you like the 7, its look is likely to be replicated throughout the BMW catalog.

What may not appear lower down the pecking order, though, is the 7 Series' innovative iDrive system —immediately obvious from its controller between the two front seats, where the gear selector should be, and its attendant display unit.

IDrive adapts to human needs, not vice versa. Distractions caused by control functions and operations are minimized, allowing full concentration on the road, while giving access to almost every control in one place.

The 7 Series is BMW's luxury model, technically excellent, like its three illustrious forebears. It has generous load space—four golf bags can be laid out crossways in the trunk. And yet, from a design viewpoint, is its compromise in rear-end design fluidity really worth this small degree of extra functionality? If you had asked the critics at the car's unveiling at Frankfurt in 2001, the answer would have been a reluctant "No."

BMW CS1

Design	Chris Bangle
Engine	1.8 in-line 4
Power	86 kW (115 bhp) @ 5500 rpm
Torque	175 Nm (129 lb ft) @ 3750 rpm
Gearbox	Sequential manual
Installation	Front-engine/rear-wheel drive
Front suspension	MacPherson strut
Rear suspension	Independent, with double control arm
Brakes front/rear	Discs/discs
Front tires	215/45ZR18
Rear tires	235/40ZR18
Length	4200 mm (165.4 ins)
Width	1650 mm (65 ins)
Height	1330 mm (52.4 ins)
Wheelbase	2600 mm (102.4 ins)
Track front/rear	1430/1460 mm (56.3/57.5 ins)
Curb weight	1500 kg (3308 lb)

The BMW CS1 concept offers the first glimpse of what is likely to become the new small BMW of the future—that is, a car that fits below the 3 Series Compact in the brand's well-honed model pecking order. This four-seater convertible flaunts a bold, independent character that remains clearly a BMW, with sporty, box-like proportions, but uses new surface forms and lines to create a distinctive look.

The most prominent features of the CS1 are its drooping sills, and the sharp shoulder form that runs around the back of the car. The wheels, positioned at each corner, are emphasized by strong wheel arches. The side surfaces flow organically from concave to convex forms, matching the strong feature lines and wheel arches. Chris Bangle's approach to body design was first seen on the X-Coupé concept in 2001.

The interior is minimalist, taking inspiration from fashion and modern architecture. The result is a fresh space that uses both classic leathers and modern materials. An aluminum cross-beam dominates the dash, and supports free-standing controls and instruments similar to those used on BMW motorcycles. These repeat the discreet light-ring design of the headlamps. The seat bases, head-rests, and steering wheel are finished in soft Nappa leather; the side panels of the seats are subject to greater loads, and so are covered in a tougher Nubuk leather. The BMW iDrive system, following its production debut on the new 7 Series (see p. 26) is fitted, to offer access to functions and information as the driver needs them.

BMW should be commended for, and proud of, the CS1. It is a vision of a new car that looks forward for inspiration at a time when so many manufacturers are delving into the past for something fresh. However, this striking new car must attract support from younger buyers to be a success when we see the production version, which will be soon.

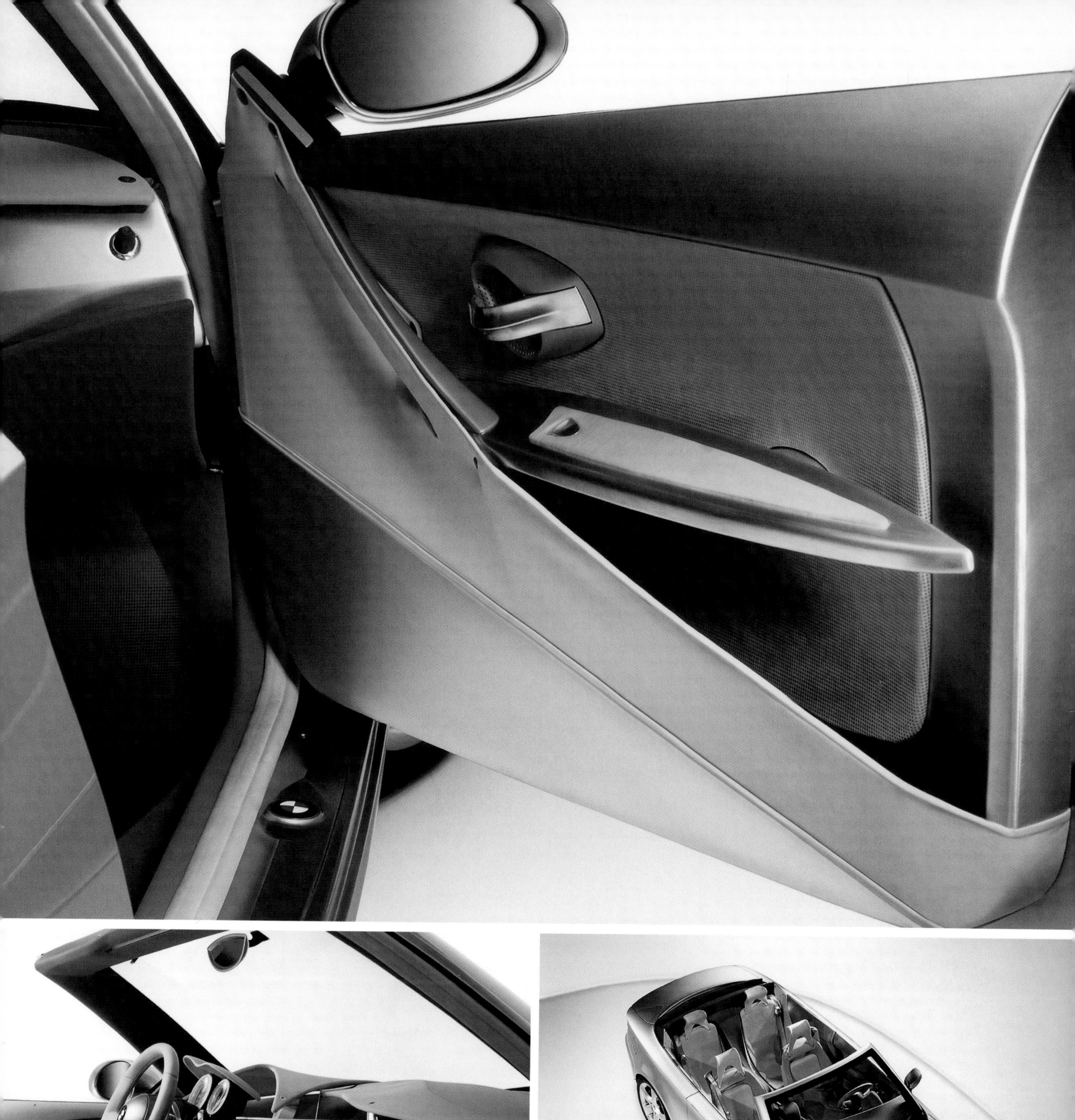

Cadillac Cien

Design	Simon Cox
Engine	7.5 V12
Power	559 kW (750 bhp)
Torque	610 Nm (450 lb ft)
Gearbox	6-speed semi-automatic
Installation	Mid-engine/rear-wheel drive
Front suspension	Double wishbone
Rear suspension	Double wishbone
Brakes front/rear	Discs/discs
Front tires	245/35R19
Rear tires	335/30R21
Length	4457 mm (175.5 ins)
Width	1975 mm (77.8 ins)
Height	1170 mm (46.1 ins)
Wheelbase	2750 mm (108.3 ins)
Track front/rear	1726/1632 mm (68/64.3 ins)

There's nothing like a birthday for a no-holds-barred celebration, and Cadillac's arrival at its centenary in 2002 is a great excuse for a party. Instead of a cake, however, General Motors' venerable luxury division presents us with the Cien. The word is Spanish for "one hundred," but the Cien is like no roadgoing Cadillac the world has seen before: it is a V12-powered, mid-engine, two-seat supercar.

However, when you recall the geometric shapes, dramatic proportions, and ultra-clean surfaces of Cadillac's other recent showstoppers, such as the Evoq, Imaj, and Vizón, you realize that the car is actually an extension of its general design philosophy, albeit lower, and sleeker.

Vertical headlamps sit alongside its dominating grille, with trapezoidal air inlets molded into the body. A feature line runs the length of the body, and another intersects it, sweeping from the tail through the sail panel to the front of the car, giving the Cien its visual drama. Cadillac even claims the razor-edged design was inspired by the latest F-22 Stealth fighter aircraft; you have to believe it.

At the rear, the Cien hints at Cadillac's flamboyant 1950s past with its fin-like vertical tail lights, and a wide, high-mounted, center brake light. There is an active spoiler that adjusts automatically, depending on the car's speed, while the roof has a removable targa panel whose blue glass apes sports-performance eyewear. The Cien's interior features a digital instrument display that relays all the car's functions. Dark anodized aluminum contrasts satisfyingly with leather upholstery.

The Cien is dramatic and visionary. It is noteworthy that Cadillac is celebrating its 100th anniversary by looking forward, rather than backward.

Cadillac CTS

Design	Thomas Kearns
Engine	3.2 V6 (2.6 V6 also offered)
Power	164 kW (220 bhp) @ 6000 rpm
Torque	296 Nm (218 lb ft) @ 3400 rpm
Gearbox	5-speed manual/5-speed automatic
Installation	Front-engine/rear-wheel drive
Front suspension	Independent, short/long arm coil over shock
Rear suspension	Independent, with high gas-charge shock absorbers
Brakes front/rear	Discs/discs
Front tires	225/55HR16
Rear tires	225/55HR16
Length	4829 mm (190.1 ins)
Width	1793 mm (70.6 ins)
Height	1440 mm (56.7 ins)
Wheelbase	2880 mm (113.4 ins)
Track front/rear	1524/1524 mm (60/60 ins)
Curb weight	1592 kg (3510 lb)
0–100 km/h (62 mph)	7.4 sec.
Top speed	238 km/h (148 mph)
Fuel consumption	10.7 ltr/100 km (21.5 mpg)

Evoq was Cadillac's mold-breaking design study in 1999, a chiseled and dramatic two-seat roadster. Its wide exposure was meant to rattle the public perception of General Motors' most upmarket brand as a provider of antediluvian sedans for America's portly and prosperous. Things then went eerily quiet, but now here comes the CTS, oozing Evoq character in real time, as it were.

The smallest Cadillac certainly needs something to set it apart. This is because it must replace the disastrous Catera. Launched in 1995, the Catera was little more than an Opel Omega wearing a Cadillac emblem. It was even German-built. It did no favors for Cadillac's image, and proved a sluggish seller.

The CTS sedan shares the Evoq's sheer forms, sharp edges, and crisp, intersecting lines. In profile, the feature line is the dominant theme, accentuating the stance of the vehicle. Traditional, "stacked" vertical headlamps and tail lights, last featured on 1965 Cadillacs, are back. The headlamps on the CTS are tall and thin; the design team claims they convey the high-tech image of optical instruments. They certainly create more space for the large, louvered, egg-crate grille—a revitalized key Cadillac feature. The rear wheels, with their blistered fenders, are a focal design point—like the car's sides, which tuck in slightly to enhance the flare of the car, and give it a more tailored look.

Warm, rich materials contrast within the cool, high-tech interior, which is contemporary, and reassuringly plush. Wood is used sparingly where the customer has tactile contact, such as the gear stick, the top of the steering wheel, and the door handles. The CTS steering wheel has a sporty, three-spoke design that features a center spline, and incorporates a distinctive "mouse" to control the in-car entertainment system.

This entry-level Cadillac is a vital new car for this perhaps tarnished brand. It must prove attractive enough to persuade Americans to switch from their luxury Japanese and European cars.

000·000

Cadillac XLR

Design	Tom Peters
Engine	4.6 V8
Gearbox	5-speed automatic
Installation	Front-engine/rear-wheel drive
Brakes front/rear	Discs/discs

Cadillac is modernizing. The XLR luxury sports car is the third new model in its line-up to be launched with a dramatic, new look aimed at younger buyers. Cadillac's edgy styling, the work of British design chief Simon Cox, already graces the CTS sedan and Escalade EXT 4×4, and was previewed on a luxury coupe in the Evoq concept at the Detroit auto show in 1999.

A two-seat convertible with a retractable hard-top, the XLR is the American equivalent of a Mercedes-Benz SL. It features the first application of a high-output, 4.6 ltr Northstar engine in a rear-wheel-drive chassis. Like the SL, the XLR has a fully automatic, retractable hard-top, in this case developed by Car Top Systems of Germany. This combines the driving enjoyment of a roadster with the comfort and additional security of a hard-top coupe.

The XLR differs only slightly from the Evoq concept. The overall impact of the styling is dramatic, with the sporty, wedge-shaped profile, and chiseled surfaces, giving a hard-edged, contemporary look. Most obvious differences are the shallower grille, the smaller air intake in the bumper, and the conventional-looking front fog lamps.

The body structure, well chosen for an open-air two-seater, uses light-weight, hydroformed, steel side rails, and an aluminum cockpit structure. Inside, the design of the two-seat cockpit combines high-tech modern materials with traditional wood trim.

RPM/1000
MPH

Chevrolet Bel Air

Design	Ed Welburn
Engine	3.5 in-line 5 turbo
Power	235 kW (315 bhp)
Torque	427 Nm (315 lb ft)
Gearbox	4-speed automatic
Installation	Front-engine/rear-wheel drive
Brakes front/rear	Discs/discs
Front tires	235/50VR18
Rear tires	235/50VR18
Length	4852 mm (191 ins)
Width	1820 mm (71.7 ins)
Height	1393 mm (54.8 ins)
Wheelbase	2820 mm (111 ins)
Track front/rear	1537/1537 mm (60.5/60.5 ins)

As redolent of 1950s America as juke-boxes and drive-in movies, the Chevrolet Bel Air—in three distinct and, today, highly collectible series between 1955 and 1957—is a post-war automotive icon. This, of course, makes it an ideal source to plunder for inspiration, which is exactly what General Motors' biggest division has done for its new Bel Air. It has, however, been orchestrated in a contemporary way.

The highly distinctive, "kick-up" belt line behind the doors revisits the original cars, as does the lofty driving position, which aims to recreate the excellent visibility, and ease of entry and exit, of large 1950s cars. This is, after all, more cruiser than sports car.

The overall effect is of a deliberately retro design with a clutter-free, barreled body, lovingly enhanced by chrome features such as the wheels, door mirrors, and windshield surround. The designers have recaptured some intriguing signature features from the 1950s Bel Airs to delight nostalgia-seekers, such as the traffic-light viewfinder, and the fuel filler cap incorporated into the tail light.

The Bel Air's interior positively drips with 1950s design cues. The twin-element instrument panel has typically 1950s typefaces. The gear stick is mounted on the steering column, and there are bench seats in the front and back. The dash is surprisingly Teutonic in its surface design, with flush-mounted switchgear, although careful application of chrome provides a good-quality feel. The only downside is an extensive use of red materials: the effect is overpowering.

This could have been a radical, forward-looking design with a salute to a glorious past. Instead, the Bel Air is a slice of modern-day kitsch, mixing old with new to create just another retro-styled American convertible.

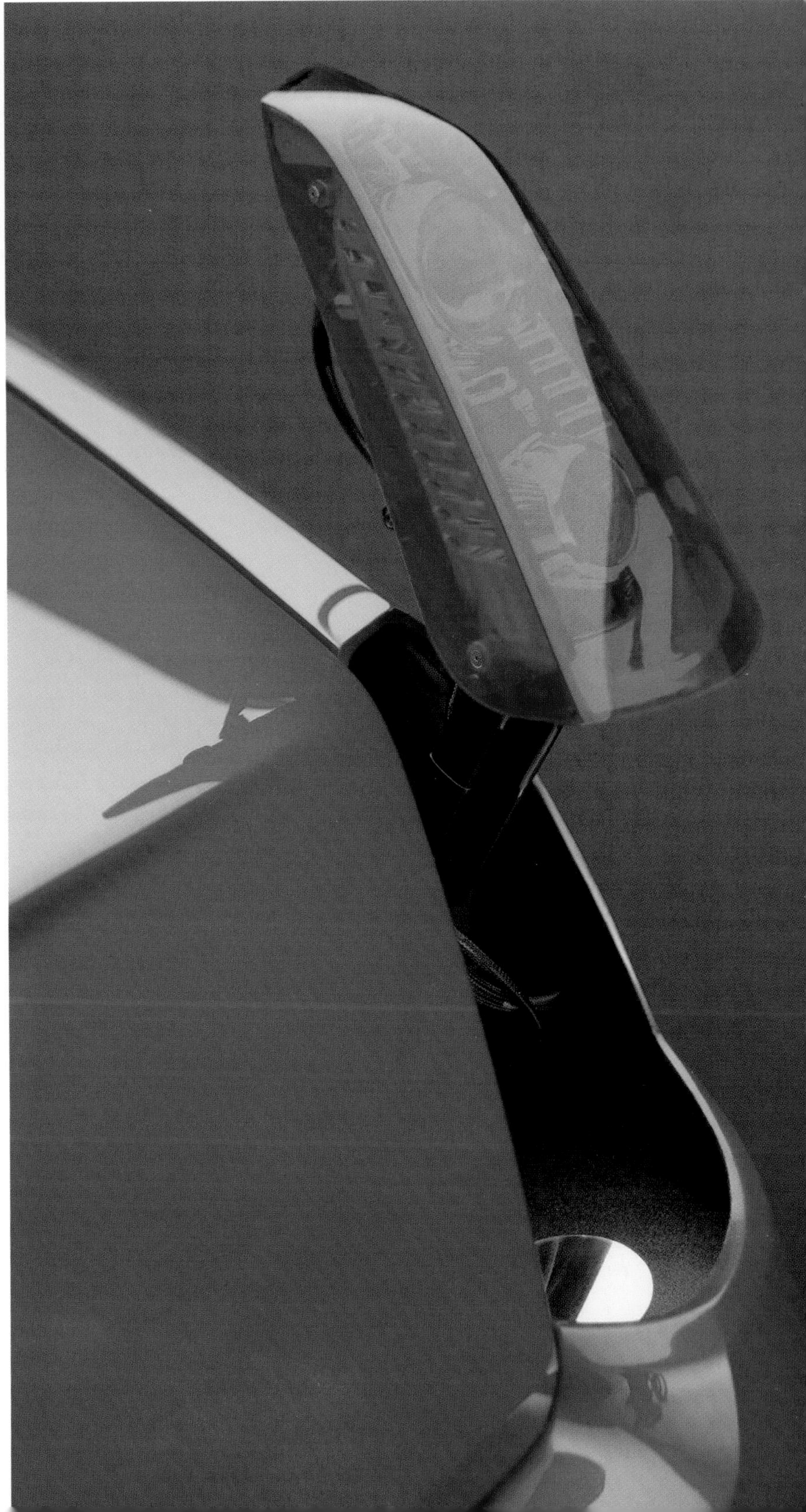

Chevrolet SSR

Design	Bill Davis
Engine	5.3 V8
Power	213 kW (285 bhp) @ 5200 rpm
Torque	441 Nm (325 lb ft) @ 4000 rpm
Gearbox	4-speed automatic
Installation	Front-engine/rear-wheel drive
Front tires	255/45R19
Rear tires	295/40R20
Length	4801 mm (189 ins)
Wheelbase	2946 mm (116 ins)

Glimpse this car looming large in your rear-view mirror, and you could be forgiven for thinking you were being pursued by a drag-racing fanatic in a customized hot rod from the 1940s. The powerfully humped hood, the two broad chrome strips topping the central grille, the flat fenders incorporating faired-in headlamps: every design detail shouts *American Graffiti*, baby boom, or the McCarthy era.

Yet, as the SSR speeds past, it becomes plain that this is not the car your father, or grandfather, might have been driving. Instead it is Chevrolet's novel 2002 take on nostalgia, turned around to draw in a modern audience. For a start, there was no such thing as a convertible pickup in the 1940s, or has not been since. This Chevrolet, however, has a stylish, solid-metal roof that miraculously hinges and folds back, Mercedes-SLK-style, under the neat load-deck cover to create an atmospheric, open-air roadster.

Just behind driver and passenger are sports-style rollover hoops. The broad, flared-out, rear wheel arches either side of the load-deck emphasize the powerful muscularity of the whole design. From behind, the SSR is again unmistakable: its load-deck cover edges up at the rear to meet the tailboard and form a spoiler, while twin exhausts poke out from under the rear apron. Most distinctive of all, though, is the full-width chrome strip that spans the car, and bisects its circular tail lights, neatly echoing a similar strip that does the same with the headlamps at the other end.

The only production car in any way close to the SSR's hot-rod concept was Chrysler's Plymouth Prowler, now out of production. That was criticized for being all show and no go—but with a large 5.3 ltr V8 engine under its hood, the Chevrolet really does have the muscle to match its looks.

Chrysler Crossfire

Design	Trevor Creed and Erik Stoddard
Engine	3.2 V6
Power	160 kW (215 bhp)
Torque	310 Nm (229 lb ft)
Gearbox	6-speed manual/5-speed automatic
Installation	Front-engine/rear-wheel drive
Front suspension	Double wishbone, coil springs
Rear suspension	Multi-link, coil springs
Brakes front/rear	Discs/discs, ABS, ESP
Front tires	225/40ZR18
Rear tires	255/35ZR19
Length	4059 mm (159.8 ins)
Width	1755 mm (69.1 ins)
Height	1300 mm (51.2 ins)
Wheelbase	2400 mm (94.5 ins)
Track front/rear	1478/1481 mm (58.2/58.3 ins)
Curb weight	1361 kg (3001 lb)
0–100 km/h (62 mph)	5.9 sec.
Top speed	238 km/h (148 mph)

The 2004 Chrysler Crossfire is a new sports coupe that combines American design with German engineering. More crucially, from a corporate viewpoint, it is also the first production car truly to pool the best elements of Mercedes-Benz and Chrysler since they were welded together as DaimlerChrysler in 1998—the biggest merger in automotive history.

The name "Crossfire," conceived during the design phase, refers to the vehicle's distinctive character line, which crosses to a negative formation from a positive one as it travels along the side of the car. Since its unveiling as a pure concept at the Detroit auto show in 2001, several things have changed. In terms of detail, gone are the large radius forms, the vertical headlamps, and the wraparound windshield with its single, vertical wiper. The rear window is now wider, and a spoiler has been added that automatically lifts at speeds over 80 km/h (50 mph). The massive 21 ins rear wheels have been replaced with 19 ins ones, to improve ride and handling, while the front wheels are also reduced, from 19 to 18 ins. The biggest difference, however, is in its engineering: a carbon-fiber body mounted on an aluminum frame is now replaced by a steel shell that uses Mercedes-Benz's CLK platform and drivetrain.

The Crossfire's crowd-pleasing overall proportions remain, as does its most distinguishing feature: the center spine that runs the length of the car from the front grille to the boat-tail, giving it a chiseled, carved appearance. This form also appears, for continuity's sake, on the door handles and exterior mirrors. The interior is a snug, two-passenger cockpit, with echoes of the exterior spine shape, which appears on the center console, gear stick, and instrument panel. Metallic highlights on the doors, steering wheel, and instrument cluster feature strongly throughout, giving the environment a technical feel.

The Crossfire is another welcome design from Chrysler, with unconventional styling that follows in the footsteps of its memorable Dodge Viper, Plymouth Prowler, and PT Cruiser models. Critics have taken issue with those cars on the grounds of engineering and quality. The Mercedes basis should calm any engineering grumbles, and quality will be looked after in German style: the Crossfire will be built by Karmann, the coachbuilder renowned for a string of classic models, including the Volkswagen Karmann-Ghia.

Chrysler Pacifica

Design	Trevor Creed
Engine	3.5 V6
Power	186 kW (250 bhp)
Gearbox	4-speed automatic
Installation	Front-engine/front- or all-wheel drive
Front suspension	Multi-link
Rear suspension	MacPherson strut
Brakes front/rear	Discs/discs
Front tires	255/50R19
Rear tires	255/50R19
Length	5052 mm (198.9 ins)
Width	1986 mm (78.2 ins)
Height	1689 mm (66.5 ins)
Wheelbase	2954 mm (116.3 ins)
Track front/rear	1676/1676 mm (66/66 ins)

With Detroit 2002 representing a post-9/11 crisis point for all the three big US manufacturers, the industry was looking to each firm's new designs for evidence of strong models that would secure a prosperous future. With Chrysler, observers needed to look no further than the Pacifica concept car to know that the trouble-torn US–German combine was back on the right track.

The sleekly shaped Pacifica artfully carves out a distinctly Chrysler identity in the fashionable field of premium SUVs. More skilfully still, smooth lines help it stay well clear of the chunky, rough-and-tough image that Jeep has nurtured so carefully in this sector. Most cleverly of all, it apes one of the most highly desirable competitors among upper-crust luxury off-roaders.

To call the Pacifica a BMW X5 with a Chrysler emblem would be a mite unfair to Trevor Creed and his design team. However, the concept undeniably evokes the aura, the proportions, and the sporty stance of the BMW. Strip away the chrome, and some of the surplus detailing, and it could almost be taken for the German product.

Everything is there, from the wagon-like sweep of the side windows and roof line, and the crisp feature line running the length of the car, to wheels that are big and impressive in a sporty, rather than an agricultural, sense. Clearly, just like the BMW, this is the shape of a four-wheel drive designed to go fast and cling to the road, rather than wade through rivers, or clamber over boulders.

Where the Pacifica departs from the BMW script is perhaps where it counts most for an American clientele: inside the cabin. Three pairs of airline-style seats give space for six. When the model goes on sale later in 2002, more conventional seating is likely to cater for seven.

In short, Voyager meets X5, but with Chrysler's keen eye for costs. The prospects look very good.

Citroën C3

Design	Donato Coco
Engine	1.6 in-line 4 (1.4 in-line 4 and 1.4 in-line 4 diesel also offered)
Power	79.8 kW (107 bhp) @ 5750 rpm
Torque	147 Nm (108 lb ft) @ 4000 rpm
Gearbox	5-speed manual
Installation	Front-engine/front-wheel drive
Front suspension	MacPherson strut, coil springs
Rear suspension	Torsion-beam axle
Brakes front/rear	Discs/discs, ABS, EBA, EBD
Front tires	185/60R15
Rear tires	185/60R15
Length	3850 mm (151.6 ins)
Width	1670 mm (65.7 ins)
Height	1520 mm (59.8 ins)
Wheelbase	2460 mm (96.9 ins)
Track front/rear	1439/1440 mm (56.6/56.7 ins)
Curb weight	1058 kg (2333 lb)
0–100 km/h (62 mph)	9.7 sec.
Top speed	192 km/h (119 mph)
Fuel consumption	6.5 ltr/100 km (43.5 mpg)
CO_2 emissions	155 g/km

The C3 is Citroën's new small car, based on the eponymous concept unveiled at the 1998 Paris auto show, to considerable acclaim. Replacing the ageing and unmemorable Saxo, the C3 has a strong and original personality. But it is not a return to the left-field approach evidenced by that legendary design classic, the 2CV.

The C3's character is conveyed through its curvaceous exterior, a cocktail of attractive surfaces and curves, while its twenty-spoke wheels and chunky, aluminum door handles express solidity. The attractive front—with body-colored, slatted bumper and grille, recessed spotlamps, and large head-lamps—looks nimble and sporty. The roof has a pronounced arch from front to rear, emphasizing its friendly, roomy, and bright interior. A small rear-quarter window adds to the feeling of interior space, while a large, electrically controlled, glass sunroof extends back over the rear seats.

The C3 features multiplex electronics, like the Citroën C5; they offer more functions with less wiring. Therefore the C3 offers lots of "intelligent" comfort and safety equipment, including parking assistance, automatic front-windshield wipers, and automatic hazard-warning lights.

The Citroën C3 is a lovely piece of chic French design, likely to appeal to men and women *en masse*. It will be a big seller, offering as it does a great alternative to the many box-like small hatchbacks.

Citroën C8

Engine	3.0 V6 (2.0 and 2.2 in-line 4, and 2.0 and 2.2 in-line 4 turbo-diesel, also offered)
Power	150 kW (201 bhp) @ 6000 rpm
Torque	285 Nm (210 lb ft) @ 3750 rpm
Installation	Front-engine/front-wheel drive
Brakes front/rear	Discs/discs, ABS, EBD, ESP
Length	4726 mm (186.1 ins)
Width	1854 mm (73 ins)
Height	1856 mm (73.1 ins)
Wheelbase	2823 mm (111.1 ins)
Track front/rear	1570/1548 mm (61.8/60.9 ins)
0–100 km/h (62 mph)	11 sec.
Top speed	205 km/h (127 mph)
Fuel consumption	11.5 ltr/100 km (20.3 mpg)
CO_2 emissions	275 g/km

Citroën's new C8 minivan replaces its Evasion/Synergie. Peugeot, Fiat, and Lancia all showed similar new minivans at the Geneva International Motor Show in 2002. Like the previous-generation models, all four are built at the Peugeot SA/Fiat Auto joint-venture plant Sevel Nord, in Lieu Saint-Amand, eastern France. Most components are common to all four cars, so they all look very similar. For instance, only body-colored side moldings, and a unique bumper and grille, differentiate the C8 from the Peugeot 807 on the outside. Collectively, they are about as close as you can get to a generic "Eurocar."

The Citroën C8 is designed as a luxurious people-carrier that combines style and modernity with practicality, comfort, and space. The new model has significantly more interior room, a huge trunk, and individual seating for up to eight occupants.

The large, bright interior is feature-rich. There are automatic, electric, sliding rear doors, and three electric, tilt/slide, glass sun-roofs with sun-blinds, one for each row of seats. The air conditioning splits the interior into four independent climate-control zones for maximum comfort. Also, just like the 807's, the distinctively styled split-level dash includes a three-sphere instrument display with an innovative "shadow-play" effect at night, making the dials very easy to read.

For safety, the Citroën C8 has a strong, reinforced body, and features Electronic Stability Program (ESP) and ABS, with electronic brake-force distribution, and emergency braking assistance, as standard. Six airbags, automatic low-tire-pressure detection, electronic parking assistance, and automatically activated headlamps are also impressive safety credentials.

CITROËN

CITROËN
CITROËN
C8

Citroën C-Crosser

Design	Donato Coco
Engine	2.0 in-line 4 diesel
Gearbox	Automatic
Installation	4-wheel drive
Front suspension	Hydractive double wishbone
Rear suspension	Hydractive double wishbone
Brakes front/rear	Discs/discs, ABS, ESP
Length	4280 mm (168.5 ins)

You'll search in vain on the C-Crosser for any of the cues that define Citroëns in the minds of the design-literate. Bear in mind, however, that the C-Crosser is intended to be a future sports utility vehicle (SUV), and this is essentially an American automotive construct.

Boldly modeled, the C-Crosser's sharply curved front end—echoed by the arc of its windshield—leads to a pronounced step at the top of the front doors, distinguishing the front and rear passenger spaces, while the high belt line, giving narrow side windows, implies an extremely solid vehicle. The C-Crosser has modular bodywork construction: it can be adapted into a rugged pickup simply by removing the rear upper section.

The C-Crosser emphasizes its interior space. It is wider than most European vehicles of this size, seating three at the front (as in the Fiat Multipla, introduced in 1999), with its passengers placed further forward, and higher, to increase interior volume, and give a panoramic view of the road. The impression of space is reinforced by the glass roof, which forms a continuous line with the windshield, bringing extra light to the rear passengers.

Drive-by-wire technology means the driver can sit on the left, in the middle, or on the right. The vehicle has four steered wheels, and an "autoactive" automatic gearbox. Ground clearance can be raised from 140 mm (5.5 ins) to 200 mm (7.9 ins), making the C-Crosser adaptable to urban or off-road use.

The C-Crosser concept, with its confident, functional styling, is a bold attempt by Citroën to design a versatile vehicle—urban people-carrier, off-road adventurer, and pickup combined—for the future. But it contains nothing of Citroën's design heritage: neither the sophistication of the DS, nor the stripped-down utility of the 2CV. The greatest irony of all is that Citroëns are not on sale in the US, which offers by far the biggest market for a vehicle like the C-Crosser.

Daewoo Kalos

Design	Giorgetto Giugiaro and DWNC Design Forum
Engine	1.6 in-line 4 (1.2 and 1.4 in-line 4 also offered)
Power	79 kW (106 bhp) @ 5600 rpm
Torque	149 Nm (110 lb ft) @ 3400 rpm
Gearbox	5-speed manual/4-speed automatic
Installation	Front-engine/front-wheel drive
Front suspension	MacPherson strut
Rear suspension	Torsion-beam axle
Brakes front/rear	Discs/drums, ABS, EBD
Front tires	175/70R13
Rear tires	175/70R13
Length	3855 mm (151.8 ins)
Width	1865 mm (73.4 ins)
Height	1495 mm (58.9 ins)
Wheelbase	2480 mm (97.6 ins)
Track front/rear	1450/1410 mm (57.1/55.5 ins)
Curb weight	1064 kg (2346 lb)
Top speed	188 km/h (117 mph)

Kalos means "beautiful" in ancient Greek, so the name of this new model is an attempt by Daewoo to leave its styling history behind. The Korean company established itself by churning out redundant General Motors models. The progress made subsequently by the Italdesign-styled Matiz economy car, and Leganza executive sedan, has been marred by the company's bankruptcy, its fugitive founder, and the lengthy search for a new owner, which has turned out to be General Motors.

A clean slate, however, means a new start, so Daewoo has gone for a compact mix of small car and MPV for the styling of the Kalos, the concept of which was first revealed at the Paris auto show in September 2000. The 2001 production model shown here is improved, friendlier, with more body color, and larger headlamps, but still—alas—awkward, angular, and fussily detailed.

The wide grille with its single, chrome-plated, horizontal slat, the geometrically cut headlamps, the flush, narrow-aperture fog lamps, and the flared fenders give it an aggressive stance. Rounded, geometric styling defines the Kalos's rear window and tail lamps. At the side, spartan lower bodywork maintains the dynamic accent, while the upper bodywork gradually narrows at the belt line, starting midway through the rear door. The visual dynamics continue on the upper part of the body, with an arching roof line and rising lower window edge, accentuated by a sharply flared line. Other sporty accents include minimal overhangs, low-profile, oblong door mirrors, flared door handles, two-tone paint, and five-spoke alloy wheels.

Simplicity and functionality are key to the Kalos's interior design. An uncluttered cockpit, and circular shapes for instruments, vents, door handles, and controls, reinforce the former attribute, while many comfort, convenience, and storage features emphasize the latter.

The Kalos is certainly different from other Daewoos. Unfortunately, from a design perspective, it is not an improvement. It is the first Daewoo from a new mold, but its well-proportioned exterior is also tortured, and some buyers will not be persuaded by it.

KALOS

KALOS

Daihatsu Copen

Design	Daihatsu design center
Engine	0.6 in-line 4 turbo
Installation	Front-engine/front-wheel drive
Length	3395 mm (133.7 ins)
Width	1475 mm (58.1 ins)
Height	1260 mm (49.6 ins)
Wheelbase	2230 mm (87.8 ins)
Track front/rear	1300/1280 mm (51.2/50.4 ins)

Although Daihatsu has no sports-car heritage to draw on, it has created a gem. The Copen roadster—its name a portmanteau of "compact" and "open"—looks great, and incorporates a neat folding hard-top, a feature increasingly found on larger, more upmarket convertibles. Its attractive exterior, with oval front and rear lamps, has a gamine and friendly appearance; its racy, six-spoke wheels and curvaceous silhouette imitate the larger Audi TT.

Could the Copen be the car that makes the Daihatsu brand desirable for younger buyers? It looks extremely credible as a production possibility, with the potential to emulate the TT's sales success. Let's hope Daihatsu can be a little bolder than usual.

That all-important hard-top is semi-automatic, meaning that it needs to be unlocked manually before the electrical system takes over. Obviously, when up it offers far better resistance toward weather and vandals than a fabric roof, but, because the Copen is so elfin, its top, when folded, devours any potential luggage space. But this car is all about driving pleasure, so perhaps the compromise is justified.

As is often the case with convertibles, the Copen looks better with the top down, when the Porsche-style, bright-red seats, and chrome roll bars protruding from the cabin—exuding sportiness—are on show to everyone.

Lilliputian Japanese sports cars, like the 1991 Honda Beat, and 1993 Suzuki Cappuccino, come along from time to time, but have tended to be passing fads, despite huge devotion from their owners. The Copen is the sort of model Daihatsu could use to capitalize on this goodwill, and make this niche market truly its own.

Copen

Daihatsu FF Ultra Space

Design	Daihatsu design center
Engine	0.6 in-line 3
Installation	Front-engine/front-wheel drive
Length	3395 mm (133.7 ins)
Width	1475 mm (58.1 ins)
Height	1720 mm (67.7 ins)
Wheelbase	2410 mm (94.9 ins)

Daihatsu's latest take on the compact, four-seat city car is this bright and roomy concept. The Japanese company already offers a production model in this class—the Move—but where that is severe and straight-lined, albeit space-efficient, the FF Ultra Space is undulatingly contemporary. Sculpted surfaces above the front and rear wheels merge toward the center of the doors, creating an attractive space-wagon style with a hint of the Smart City Coupé's radicalness.

The FF Ultra Space looks particularly spacious thanks to the front side windows, arched and sitting higher than the rear ones. The whole upper architecture is glazed for excellent driver visibility. Rear-seat access benefits from the absence of a pillar between the front and back doors, and from the usefulness of the sliding rear door.

The interior is curvaceous, and brightly colored. Meters are grouped at the center of the instrument panel, and interior lines spread dynamically from the center, creating a futuristic atmosphere. Triangular shapes are used throughout, echoing the Daihatsu emblem.

When the FF Ultra Space is reversed, the car's navigation system displays the scene behind the vehicle, using a wide-angle, monochrome CCD camera. When the car is parked, its front seat can revolve 180 degrees, to create a meeting space, or an impromptu dining area.

The FF Ultra Space is a neat-looking concept that would sit comfortably in a twenty-first-century city environment, as well as doing its part to boost the sometimes overlooked Daihatsu brand. Regrettably, despite the car featuring production-ready details, Daihatsu maintains that there are no current plans to build it.

Daihatsu Muse

Design	Daihatsu design center
Engine	0.6 in-line 3
Installation	Front-engine/front-wheel drive
Length	3395 mm (133.7 ins)
Width	1475 mm (58.1 ins)
Height	1510 mm (59.4 ins)
Wheelbase	2360 mm (92.9 ins)

It is impossible to attend an auto show in Tokyo without being incredulous at the sheer number of concept vehicles on display from Japanese manufacturers. It is almost as if they feel obliged to present a range of concept cars as large as their production palettes. There will be the production-ready concept, the wild concept, the sporty concept, and the retro concept. Most are never seen again. The Daihatsu Muse is likely to be one of these, an insignificant design that simply helped make up the numbers at Tokyo in October 2001.

Its 1960s-inspired, box-like shape, long cabin, and thick doors are supposed to impart a feeling of security. Detailing, meanwhile, is modern, with chunky, horizontal door handles, clear headlamps, aluminum wheels, and turn signals mounted in the door mirrors.

No doubt this would be a practical car if it were ever offered to customers. The windshield and side windows are upright, and the seats are positioned high, so visibility is excellent. The short wheel overhangs would make parking easy. And the low, flat floor, together with doors opening to ninety degrees, would make entry and exit effortless. The simple interior is dominated by an instrument panel fitted with a 265 mm (10.4 ins) monitor sitting on the center of the dash. Drive-by-wire technology would, apparently, permit hands-free gear-shifting.

As ever, Daihatsu's most recent clutch of concept cars varies greatly in style and desirability. The Muse—with its trapezoid shape, and odd-looking front side window—is a boxy miss, while the Copen is a curvy hit. Retro buffs might like the Muse, but it is a design irrelevance.

MUSE

Daihatsu U4B

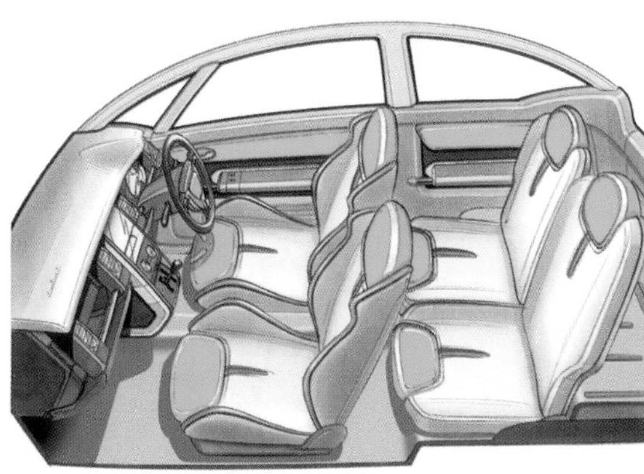

Design	Daihatsu design center
Engine	0.6 in-line 3
Installation	Front-engine/4-wheel drive
Length	3395 mm (133.7 ins)
Width	1475 mm (58.1 ins)
Height	1570 mm (61.8 ins)
Wheelbase	2360 mm (92.9 ins)

For once, Daihatsu's designers have created a stylish minicar with the U4B (shorthand for Urban 4×4 Buggy). This is Daihatsu's attempt to interest fashion-conscious buyers in its products, using a mix of city-car looks and off-road abilities. The closest it has come to such a car before is the tall and narrow Terios baby off-roader, which suggests a half-size Land Rover Discovery.

An "urban 4×4" is a contradiction in terms, of course, but the market is dictating a growing demand for off-road looks, and some actual off-road capability, in everyday cars. There is a clear reference to the Volkswagen Beetle in the arched overall shape, bulging front wheel arches, oval headlamps, and round door mirrors. Rugged-looking wheels, and gray plastic body trim, hint at off-road aspirations.

The glass roof has five panels, which can be opened to allow more air into the cabin. Blistered wheel arches emphasize the amount of available wheel travel—very necessary in any off-road vehicle. The fully glazed upper structure follows a common trend in many recent concept cars, while the interior is designed to be fun, with some aluminum technical detailing, consistent with the off-road imagery.

Daihatsu has designed a funky-looking, sporty car with an interesting off-road element that is sure to appeal to younger people. Only the blinkered outlook of its marketing department stands between the U4B and many potential buyers.

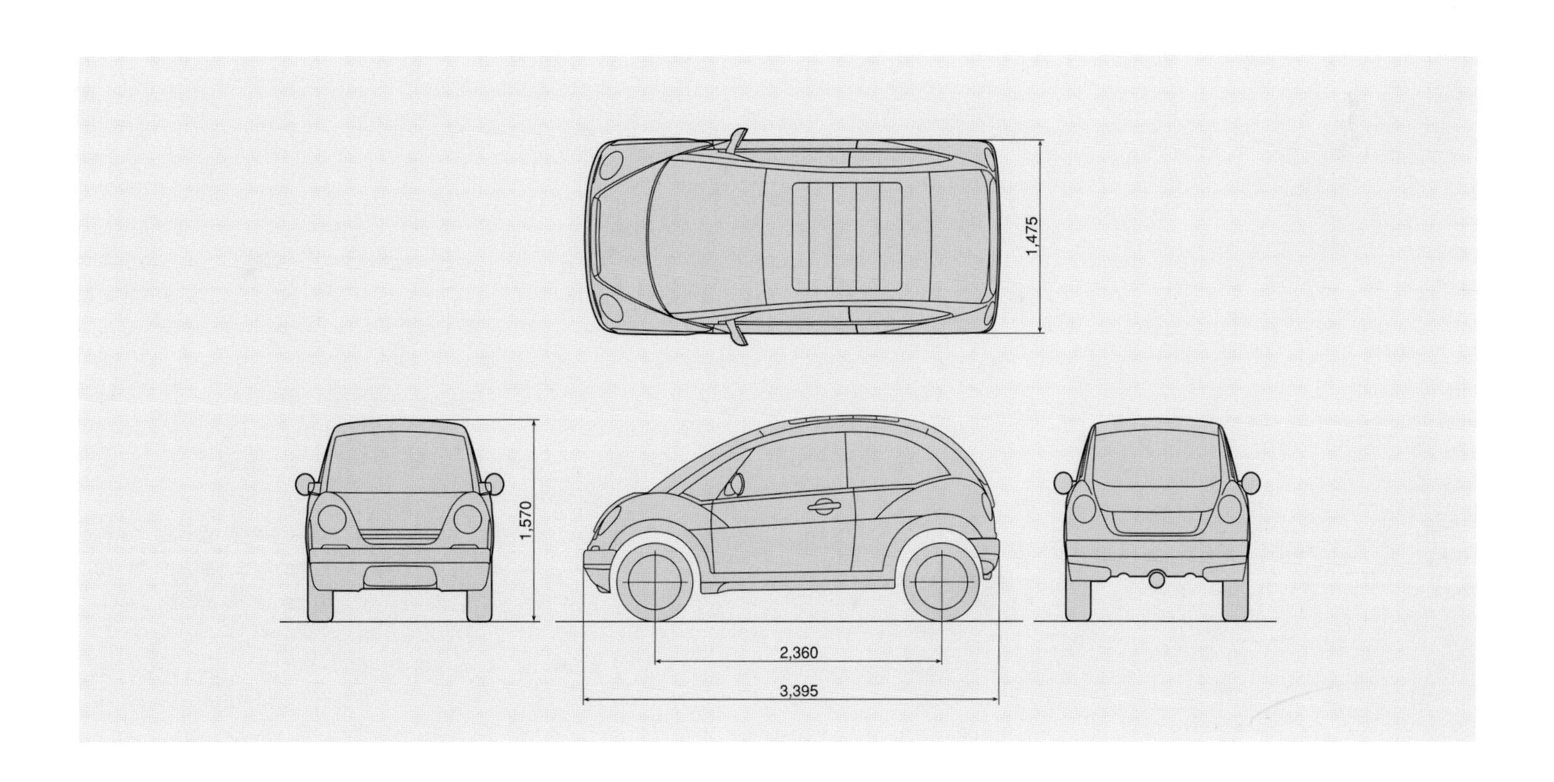
1,475
1,570
2,360
3,395

Daihatsu UFE

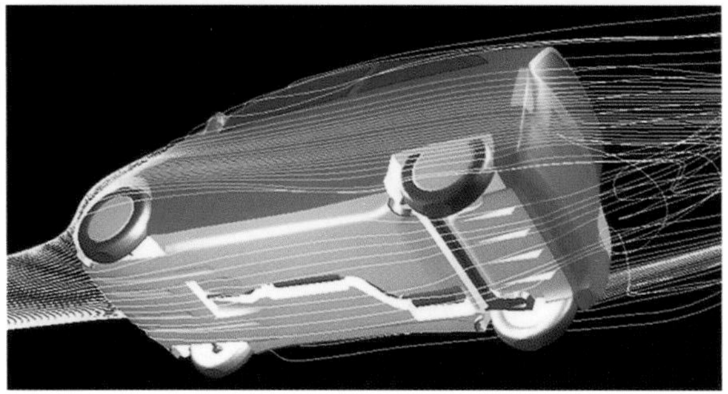

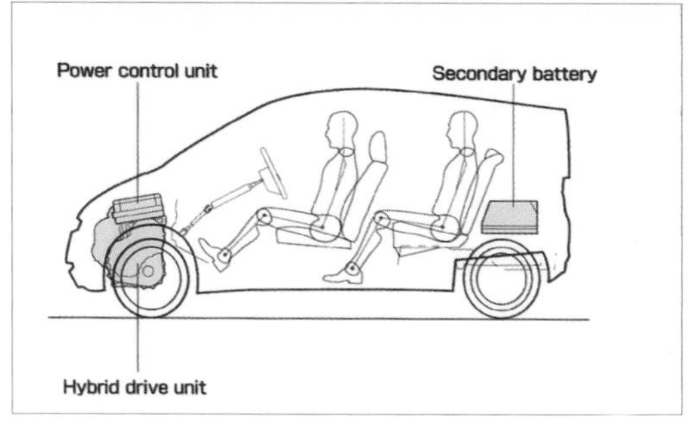

Design	Daihatsu design center
Engine	0.6 in-line 3, plus electric motor
Installation	Front-engine/front-wheel drive
Front suspension	MacPherson strut with coil springs
Rear suspension	Trailing arm with torsion beam
Brakes front/rear	Discs/drums
Front tires	135/80R13
Rear tires	135/80R13
Length	3395 mm (133.7 ins)
Width	1475 mm (58.1 ins)
Height	1475 mm (58.1 ins)
Wheelbase	2380 mm (93.7 ins)
Track front/rear	1300/1260 mm (51.2/49.6 ins)
Curb weight	630 kg (1389 lb)
Fuel consumption	1.82 ltr/100 km (153 mpg)

Daihatsu's UFE (Ultra Fuel Economy) has been built to achieve outstanding fuel economy, and also to offer some advanced safety technologies. With a weight of 630 kg (1389 lb), the UFE can cover 55 km (34 miles) on just one liter of fuel.

A composite plastic body with a largely aluminum platform is the reason for this feather-lightness. The smooth, uncluttered body resembles a cut-off teardrop, but, more importantly, it gives a drag co-efficient (Cd) of a wind-cheating 0.25. To achieve this, the rear wheels are enclosed, the air intakes and door mirrors are reduced in size, and only the windshield wipers—practical, if profoundly un-aerodynamic—are still prominent.

The UFE's hybrid power unit is fundamental to its fuel-conserving concept. It features a direct-injection gasoline engine, and an electric motor that switches on or off automatically, depending on outside factors such as road speed or acceleration. A regenerative braking system has also been fitted; this acts as a generator during braking, converting kinetic energy into electricity, which is then stored in the battery for re-use. The drivetrain broadly follows that of the award-winning Toyota Prius hybrid production car. Toyota, of course, is Daihatsu's controlling shareholder.

The UFE also has plenty of technically interesting safety elements, such as adaptive cruise control, collision avoidance, lane following, and speed control for cornering. A night-vision system detects objects beyond the range of the headlamps, whose beam direction changes with the steering.

The UFE's streamlined design is echoed in the interior, where a winged instrument panel wraps around the door panels to create a continuous band of color. The aerodynamic shape, however, is an unfortunate one—you could never describe the UFE as an attractive-looking car, no matter how infrequently you would need to return to the gas station.

UFE
UFE HYBRID SYSTEM

Dodge M80

Design	John Opfer
Engine	3.7 V6
Power	157 kW (210 bhp)
Torque	319 Nm (235 lb ft)
Gearbox	5-speed manual
Installation	Front-engine/4-wheel drive
Front suspension	Independent, with upper and lower A-arms
Rear suspension	Live axle
Brakes front/rear	Discs/discs
Front tires	265/50R20
Rear tires	265/50R20
Length	4229 mm (166.5 ins)
Width	1631 mm (64.2 ins)
Height	1687 mm (66.4 ins)
Wheelbase	2845 mm (112 ins)
Track front/rear	1537/1537 mm (60.5/60.5 ins)
Curb weight	1134 kg (2501 lb)
0–100 km/h (62 mph)	8 sec.
Top speed	160 km/h (100 mph)

The children of America's post-war baby boomers are the target group of the Dodge M80, a retro-styled pickup truck. Dubbed "millennials" by Dodge, part of the German-controlled DaimlerChrysler group, the members of the anticipated market are under twenty-four years old—hence the brightly colored yellow paint, which has emerged as a common theme for trucks in 2002 (see also the Ford F-350 Tonka on p. 70).

Although modern in execution, the M80's shape draws its influence from 1940s Dodge pickups, such as the TC, from 1939, and WC, from 1941. Today, an M80 is more likely to be hauling a snowboard and mountain bike than a pile of lumber, so the Dodge concept is designed around a low-cost, four-wheel-drive Dodge Dakota pickup chassis, a decision that boosts the M80's chances of production.

Aggression is added to the design through huge wheel arches, which emphasize the large wheels, and lend credibility to its off-road ability. Other classic design references are circular door mirrors, round headlamps, and slatted vents in the front fenders. These are matched to a powerful aluminum front grille with a modern, technical look.

To be credible, any pickup still has to be practical, so the load bay is protected by a plastic liner, and incorporates tie-down cleats for bikes and surfboards. A sign that the M80 is more lifestyle statement than practical is the 1.5 m (4.8 ft) load bay. Usually an American pickup has a 2.5 m (8.1 ft) bed, big enough to carry a builder's standard-size sheet of plasterboard or plywood.

"No frills" is a good description for the interior, with body-colored panels where soft trim would normally soften the cabin's ambience. Extensive use of satin-silver on the center console, steering wheel, and doors adds a technical feel, while the use of water-repellent Neoprene-look trim boosts the rugged design.

MPH

Dodge Razor

Design	Akino Tsuchiya and Kevin Verduyn
Engine	2.4 in-line 4 turbo
Power	187 kW (250 bhp)
Torque	312 Nm (230 lb ft)
Gearbox	6-speed manual
Installation	Front-engine/rear-wheel drive
Front suspension	MacPherson strut
Rear suspension	Multi-link independent, with coil springs
Brakes front/rear	Discs/discs
Front tires	185/50R19
Rear tires	185/50R20
Length	3754 mm (147.8 ins)
Width	1748 mm (68.8 ins)
Height	1214 mm (47.8 ins)
Wheelbase	2489 mm (98 ins)
Track front/rear	1473/1524 mm (58/60 ins)
Curb weight	1134 kg (2501 lb)
0–100 km/h (62 mph)	6 sec.
Top speed	225 km/h (140 mph)

Simple British sports cars of the 1960s were the inspiration for the Dodge Razor, one of three concepts from the DaimlerChrysler subsidiary at the 2002 Detroit auto show.

A lightweight two-seater that is more compact than its main competitor, the Mazda MX-5, the Razor could retail at just $14,500 in North America, a keen price aimed at college kids. Dodge's focus on this youthful market also explains the choice of name, which is borrowed from the trendsetting Razor USA brand.

Dodge's Razor has a simple shape with a long hood, wide stance, and narrow "greenhouse"—all features that breathe emotion into the design—and, in profile, the shape is similar to the earliest Porsche 911. The only exterior ornamentation is the chrome bumpers, the racing-car-style, forged-aluminum fuel cap, the classic door handles and mirrors, and the jewel-like lamp design.

Like the Dodge M80 (see p. 62), the Razor concept employs a no-frills interior, although the theme is race-car-inspired, with lightweight bucket seats and four-point seat belts. Body-colored interior panels and extruded aluminum features reinforce the sporty cabin feel.

More unusual is the integrated tachometer and speedometer, with its conventional, analog tach needle matched to a futuristic, digital read-out for speed—an idea already in use on the Porsche Boxster.

Under the Razor's skin are mechanical parts borrowed from other models, but the all-important chassis platform is unique, which sadly makes the chances of the Razor reaching production in this exact form pretty remote.

Fiat Ulysse

Design	Fiat Style Center
Engine	3.0 V6 (2.0 in-line 4 and 2.2 in-line 4 turbo-diesel also offered)
Power	150 kW (204 bhp) @ 6000 rpm
Torque	285 Nm (210 lb ft) @ 3750 rpm
Gearbox	Manual or automatic
Installation	Front-engine/front-wheel drive
Front suspension	MacPherson strut
Rear suspension	Deformable beam
Brakes front/rear	Discs/discs, ABD, ESP
Length	4720 mm (185.8 ins)
Width	1860 mm (73.2 ins)
Height	1856 mm (73.1 ins)
Wheelbase	2823 mm (111.1 ins)
Track front/rear	1570/1548 mm (61.8/60.9 ins)
0–100 km/h (62 mph)	10.2 sec.
Top speed	205 km/h (127 mph)
Fuel consumption	11.5 ltr/100 km (20.3 mpg)

The new Fiat Ulysse, another of the "Sevel" minivans produced by the Peugeot/Fiat joint venture (see Citroen C8 on p. 46), replaces the existing version in 2002. It has taut, clean lines, and architectural shapes with a strong Italian flavor. Hence the new Ulysse gets a pronounced personality, despite the fact that such a boxy profile usually presents a thorny challenge to designers.

At the front, a round shield on a slightly oval grille with a honeycomb grid dominates, together with slender, high-tech headlamps. The long body sides have crisp lines that run toward the rear. Sleek door mirrors extrapolate the line of the wings, and ultimately lead to the side turn signals. At the rear, tall tail lights frame the large tailgate.

Inside the Ulysse, a distinctive arched dash offers three dials that appear to hang below it. The center console extends toward the passenger compartment, to bring the shift and climate controls within easy reach. It can also be fitted with CONNECT, a system Fiat asserts is the most advanced integrated "infotelematic" device on the market. It acts as an on-board assistant that can be asked to make a phone call, put on music, or take a memo, and is operated by remote control.

The seats are less rounded than before, which is made possible by a new manufacturing technique that achieves sharper angles while minimizing the number of seams.

The target customers for the new Ulysse are people with an energetic, fun lifestyle—Fiat calls it "a life in primary colors," whatever that means. The Ulysse actually suits large families, or occasional cargo carriers.

FIAT Ulysse

FIAT
AIRBAG

Fioravanti Yak

Design	Fioravanti
Engine	V8
Installation	Front-engine/4-wheel drive
Length	4500 mm (177.2 ins)
Width	1910 mm (75.2 ins)
Height	1680 mm (66.1 ins)
Wheelbase	2800 mm (110.2 ins)
Track front/rear	1565/1565 mm (61.6/61.6 ins)

The Yak, by independent Turin design consultancy Fioravanti, is a "crossover" concept that combines the features of a station wagon with an SUV.

The front of the Yak sports several air inlets and outlets, plus an eye-catching array of lamps. The lamps incorporate 50-lumen LEDs with special lenses, controlled by software to obtain different light beams. For example, you can lengthen the depth of the beam at fast speeds on straight roads, widen the low beam at low speeds in cities, or adjust low/high beams on mountain roads.

The Yak's structure is based on a cross-ring element, a sort of central rollover bar. It incorporates another unusual feature: side-window wipers for all four doors, which improves visibility by 75%. The side surfaces draw the eye to the door pillar that houses the wiper mechanisms.

Inside, the dash has voice-activated controls. The "infotainment" system is mouse-operated, and can also be accessed from the rear seat by a control mounted on the rollover bar structure.

Fioravanti's Yak has a patented system for semi-automatic seat belts. A sensor in the seat activates the armrests a few seconds after the passengers sit down, placing the top parts of the four-point belts over their shoulders, and leaving only the driver to fasten a central buckle.

This new design style, loaded with innovative and imaginative ideas, should be lauded. Fioravanti is pushing out the boundaries of car design—even if this is not its most attractive concept—and that deserves some congratulations.

Ford F-350 Tonka

Design	Patrick Schiavone
Engine	6.0 V8 turbo-diesel
Power	261 kW (350 bhp) @ 3300 rpm
Torque	813 Nm (600 lb ft) @ 2000 rpm
Gearbox	5-speed automatic
Installation	Front-engine/rear- or 4-wheel drive
Front suspension	Live axle with air springs
Rear suspension	Live axle with air springs
Brakes front/rear	Discs/discs, ABS
Front tires	315/60R22
Rear tires	315/60R22, dual rear wheels
Length	6147 mm (242 ins)
Width	2362 mm (93 ins)
Height	2134 mm (84 ins)
Wheelbase	4191 mm (165 ins)
Track front/rear	1880/2032 mm (74/80 ins)
0–100 km/h (62 mph)	7 sec.
Top speed	201 km/h (125 mph)

The Ford F-350 Tonka, 2.1 m (7 ft) high, 2.3 m (7.8 ft) wide, and 6.1 m (20.2 ft) long, comes across not so much as a larger-than-life caricature of the archetypal all-muscle pickup truck, but as a barely scaled-down version of an interstate-pounding Mack truck.

Huge 19 ins wheels, double at the rear, give the F-350's structure massive ground clearance, and more than a hint of the customized car-crushers that inhabit country-fair arenas. The substantial clearance between the trademark, circular wheel arches and the tires suggests off-road prowess. Heavy use of chrome dominates the smooth exterior style, again in the manner of classic American truck rigs, and the raised hood is a strong statement of the big 6 ltr V8 diesel that lies beneath.

Inside the double cab, the theme switches from chrome and yellow to blue—in the form of leather and soft furnishings—and polished aluminum for the structure and highlights. The vastness of the cab is further emphasized by the absence of a central pillar, the rear doors being hinged at the back to provide wide-open access to the space inside.

The Tonka driver's seat is more akin to a commander's post. It features an all-terrain, shock-resistant suspension system for greater comfort on punishing roads. The interior style is technical in its impression, with precise detailing, and a preference for hard, straight edges, and angular shapes. The extensive use of aluminum alloy and stainless-steel fittings helps to echo the strength and durability conveyed by the exterior features.

The Ford F-350 Tonka may appear a brutish and grossly exaggerated piece of design, with little relevance to contemporary road conditions, and no apparent purpose other than to shock and impress. But that also makes it perhaps the perfect concept car for the Auto City's biggest show.

Ford Fiesta

Design	Chris Bird
Engine	1.6 in-line 4 (1.3 and 1.4 in-line 4, and 1.4 in-line 4 diesel, also offered)
Power	74 kW (99 bhp) @ 6000 rpm
Torque	143 Nm (105 lb ft) @ 4000 rpm
Gearbox	5-speed manual
Installation	Front-engine/front-wheel drive
Front suspension	MacPherson strut, coil springs
Rear suspension	Semi-independent twist beam, coil springs
Brakes front/rear	Discs/drums, ABS, EBD
Front tires	175/65R14
Rear tires	175/65R14
Length	3917 mm (154.2 ins)
Width	1683 mm (66.3 ins)
Height	1417 mm (55.8 ins)
Wheelbase	2487 mm (97.9 ins)
Track front/rear	1477/1444 mm (58.1/56.9 ins)
Curb weight	1170 kg (2580 lb)
0–100 km/h (62 mph)	10.8 sec.
Top speed	185 km/h (115 mph)
Fuel consumption	6.6 ltr/100 km (35.5 mpg)
CO_2 emissions	158 g/km

The new Fiesta was launched in September 2001, a critical time for Ford: its president Jacques Nasser had just been ousted, it was scaling down production worldwide, and, on top of all this, it was recovering from the Firestone tire crisis. Perhaps that is why the Fiesta appears to play safe when compared to such adventurous recent Fords as the Ka and Focus.

The company's design aim for the new Fiesta was to reflect its trademark "edge" styling philosophy, while creating a characterful small car with solidity and space. The large, friendly-looking headlamps follow the curves of the front corners, while the smaller, bumper-mounted driving lamps give an impression of zest.

The sharp feature line running along the side of the car drops down toward the front, enhancing its dynamic poise. The exterior is similarly proportioned to the model it replaces, so retaining its links to the Fiesta heritage—one that stretches back an astonishing twenty-five years, despite using only two basic designs. The original, square-cut, 1976 Fiesta remains one of just two Ford production models to be designed at its Ghia studios in Turin, Italy. The body has been simplified, and made more harmonious, with more volume in its surfaces. These are the design cues central to every new European Ford. The cabin area has been redesigned too: the hood is short, while the glass area is generous, and extends into a third side window, giving rear passengers a sense of greater light and space.

The dash is punctuated by large, circular air vents that add an element of fun and cheerfulness to the new Fiesta's interior. It is completely redesigned, with a new instrument binnacle fully integrated into the central console in an inverted L-shape, providing the driver with maximum control from the cockpit.

The Fiesta is clearly reminiscent of its stablemate, the Focus, if not quite so visually arresting. There is no reason why it shouldn't follow its predecessor in sharing it's elder brother's popularity.

FIESTA
K FK 6023
TDCi
40

K FK 6069

Ford Fusion

Design	Chris Bird
Engine	1.4 in-line 4 diesel (1.4 and 1.6 in-line 4 gas also offered)
Power	50 kW (67 bhp) @ 4000 rpm
Torque	160 Nm (117 lb ft) @ 1750 rpm
Gearbox	Automatic
Installation	Front-engine/front-wheel drive
Front suspension	MacPherson strut, coil springs
Rear suspension	Semi-independent and twist beam
Brakes front/rear	Discs/drums
Front tires	195/60R15
Rear tires	195/60R15
Length	4020 mm (158.3 ins)
Width	1708 mm (67.2 ins)
Height	1503 mm (59.2 ins)
Wheelbase	2488 mm (98 ins)
Track front/rear	1472/1435 mm (58/56.5 ins)
0–100 km/h (62 mph)	15.5 sec.
Top speed	158 km/h (98 mph)
Fuel consumption	4.4 ltr/100 km (53.2 mpg)
CO_2 emissions	119 g/km

If you feel there is something rather conservative, almost bland, about this Ford concept car, then you are on the mark. Its neat and restrained lines, first revealed at the Frankfurt Motor Show in 2001, are meant to be inoffensive: the Fusion made its production debut at the 72nd Geneva International Motor Show in 2002 and will be going on sale in 2003, so its customer-ready image is for today, rather than the future. It will be Ford's first small SUV and the embattled company is loath to scare away even one potential buyer from this profitable niche market.

This Fusion "taster" has a two-tone body, its plastic lower section providing protection for its intended, sparing off-road use. The headlamps, fog lamps, and mirror-mounted turn signals have three internal segments and metallic surrounds, all echoing one another.

Although commercial logic dictates that the showroom edition of the Fusion will be based on the Fiesta platform, the concept Fusion bristles with high technology. Ford's Durashift Electronic Select Transmission combines the benefits of automatic and manual gearboxes, with electronically activated gearshift and clutch operation to prevent sudden gear changes and undesirable shifts up and down. The new navigation system was developed in cooperation with Visteon. In addition to navigation assistance and modern Internet functions—such as downloading news, pictures, or music—the system offers real-time traffic and parking information. Also featured is a new Hella bi-Xenon headlamp system, which provides exceptional levels of visibility at night. Xenon light is two-and-a-half times as powerful as that produced by halogen, but uses less than two-thirds of the power.

The Ford Fusion concept's technologies could well be incorporated into other Ford models. The Fusion mini-SUV, however, will definitely become part of Ford's future range.

FUSION
Ford
fusion

Ford
fusion

Ford GT40

Design	Camilo Pardo
Engine	5.4 V8
Power	373 kW (500 bhp) @ 5250 rpm
Torque	678 Nm (500 lb ft) @ 3250 rpm
Gearbox	6-speed manual
Installation	Mid-engine/rear-wheel drive
Front suspension	Unequal-length control arm, push-rod/bell-crank system with longitudinal/horizontal spring-dampers
Rear suspension	Unequal-length control arm, push-rod/bell-crank system with longitudinal/horizontal spring-dampers
Brakes front/rear	Discs/discs
Front tires	245/45R18
Rear tires	285/45R19
Length	4613 mm (181.6 ins)
Width	1950 mm (76.8 ins)
Height	1106 mm (43.5 ins)
Wheelbase	2710 mm (106.7 ins)
Track front/rear	1636/1650 mm (64.4/65 ins)

The centenary of the Ford Motor Company falls in 2003. The re-creation of its most famous racing car, the 1960s Le Mans-winning GT40, is the centerpiece of the company's celebrations.

Cleared for production just two months after the concept was revealed at the 2002 Detroit auto show, the GT40 is the automotive version of the retro remake of a watch, or a toy car. It marks the second dip into Ford's design archives after the Thunderbird two-seater was re-created two years ago.

The GT40 also indicates a further strengthening of the trend to relaunch classic cars in the modern era. In fact, Ford design boss J. Mays started the trend when he was a designer at Volkswagen, where he created the new Beetle. BMW pulled a similar trick when replacing the Mini, and Volkswagen is trying to repeat its success with a remake of the Microbus.

Modernizing the GT40 won't be easy, however. The concept is bigger than the old one, including its key dimension—height. The "40" in its name refers to the original's 40 ins (just over 1 m) height—the 2002 version is 43.5 ins (1.12 m) high. Ford may even have to use the latter number, having sold the rights to the original name in the 1970s to specialized British car manufacturer Safir Engineering. A resolution to that problem is being negotiated.

Otherwise the GT40 is faithful to the sweeping styling of the 200 mph (322 km/h) original, with its long front overhang, aggressive air intakes for the radiators, narrow "greenhouse," and kicked-up aerodynamic tail. And in the tradition of championship racers, the doors are cut into the roof.

GT40

GM AUTOnomy

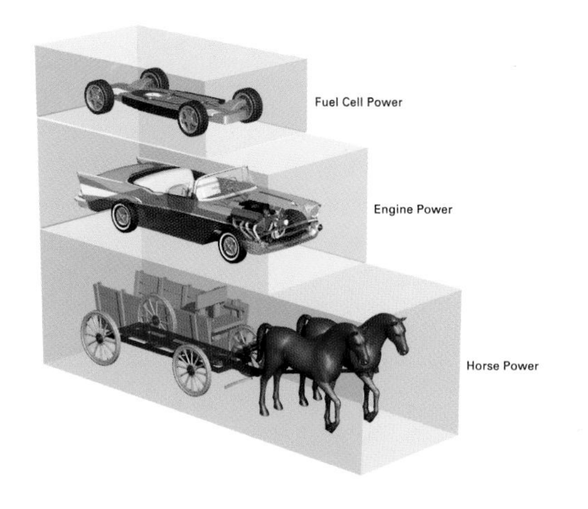

Design	Wayne Cherry
Engine	Hydrogen fuel-cell system
Front tires	195/55R22
Rear tires	195/55R22
Length	4465 mm (175.8 ins)
Width	1880 mm (74 ins)
Height	1247 mm (49.1 ins)
Wheelbase	3099 mm (122 ins)
Track front/rear	1651/1727 mm (65/68 ins)

Here is a concept car that really does explore new concepts. The mission of General Motors' AUTOnomy is to be a technology demonstrator rather than a style ambassador. Its task is to show the remarkable design and configuration horizons that can be opened up when radical new propulsion systems—in this case fuel-cell power—are designed in from the start.

So forget for the moment the sleek and futuristic two-seat coupe superstructure that General Motors claims was inspired by motorcycle and jet-fighter design; forget, too, the exterior shape, largely dictated by low-drag aerodynamics, and even the mono-form cabin with its fenders that extend outward to provide downforce and cover the suspension and wheels. These are just attention-seeking distractions, for what is truly important about the AUTOnomy is its chassis concept.

General Motors' HydroGen II fuel-cell powertrain offers designers unprecedented versatility when it comes to placing the essential mechanical elements within the vehicle. In this case, the engineers have chosen a thin, sandwich-type chassis, rather like a skateboard, to house the fuel cells and their control electronics; the Wheel drive is electric, and, because drive-by-wire control technology has replaced all the conventional mechanical controls, there is no restriction as to where the driver and the handlebar-like control module can be located.

This is a chassis concept of some significance. It raises the prospect of a standardized autonomous platform providing all propulsion and control functions, capable of being built in large numbers and at low cost. Such is the versatility of the platform that it can support a wide variety of bodies—anything from a single-seat commuter to a seven-seat minivan and, indeed, exterior shapes that we have never before imagined possible.

Drive-by-wire control and fuel-cell-propulsion technology have been shown on past concept cars, usually with only slightly modified vehicle architecture. The AUTOnomy, on the other hand, makes the most of these advances to define a completely new way of building a vehicle.

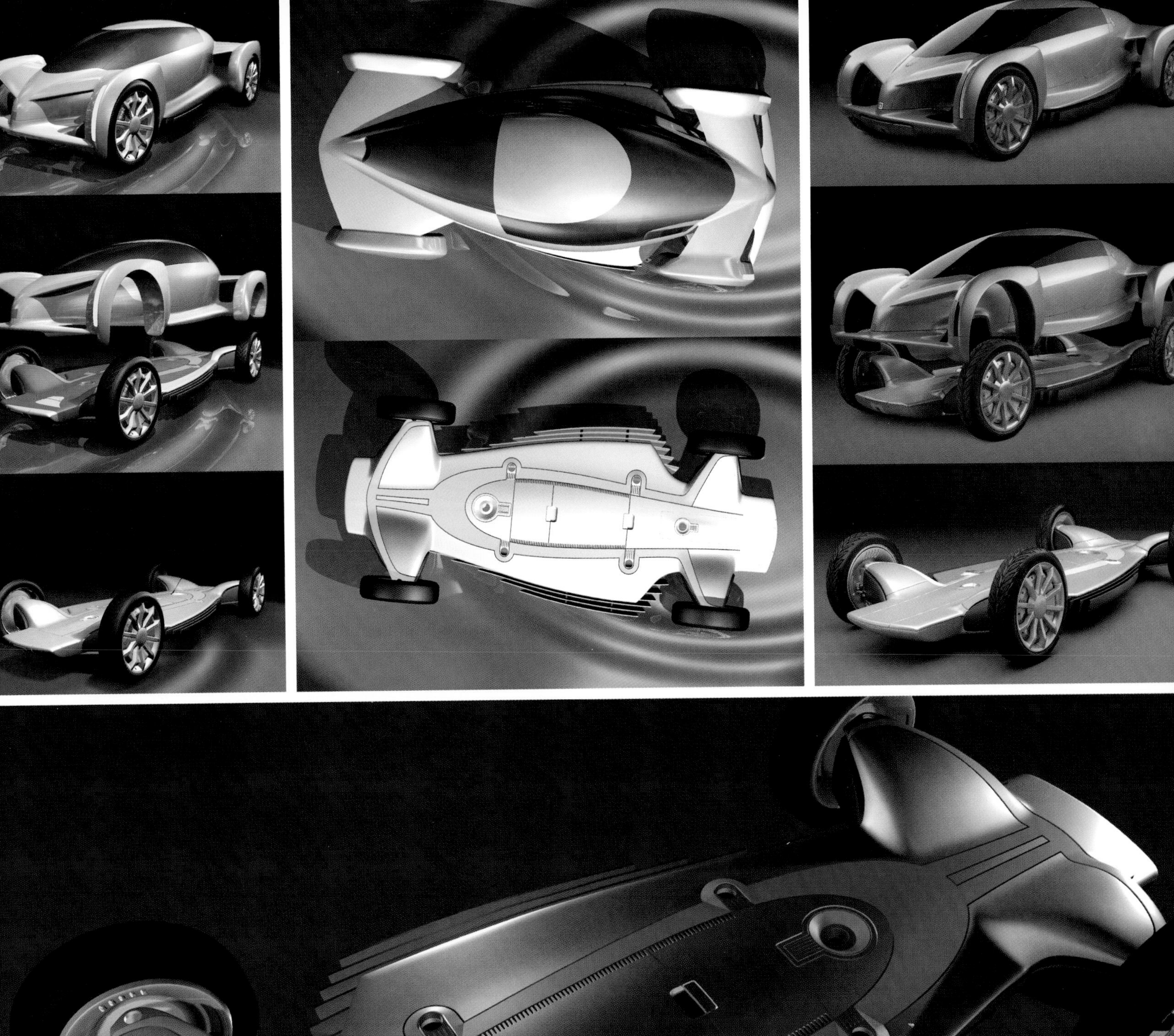
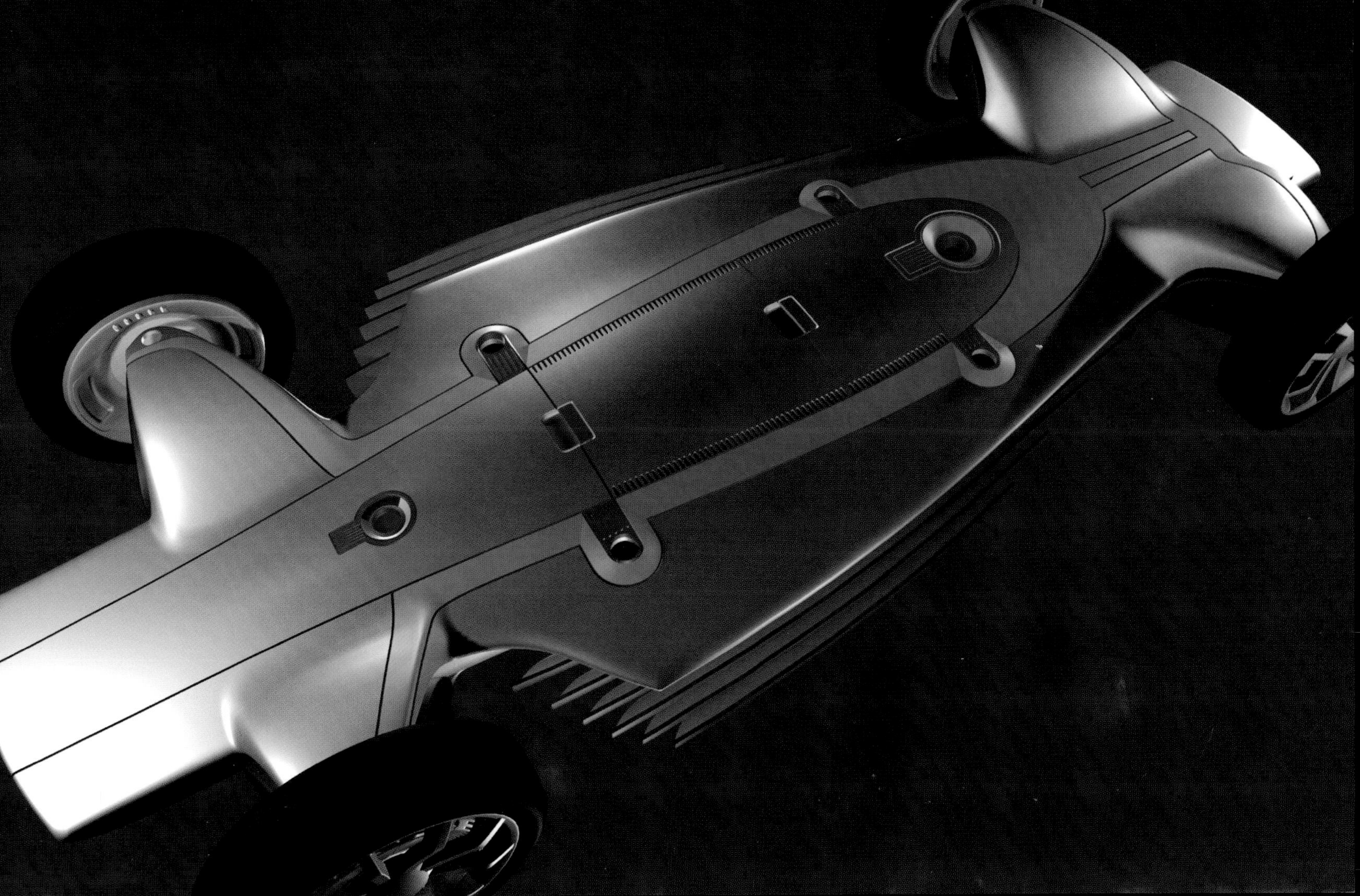

Honda Bulldog

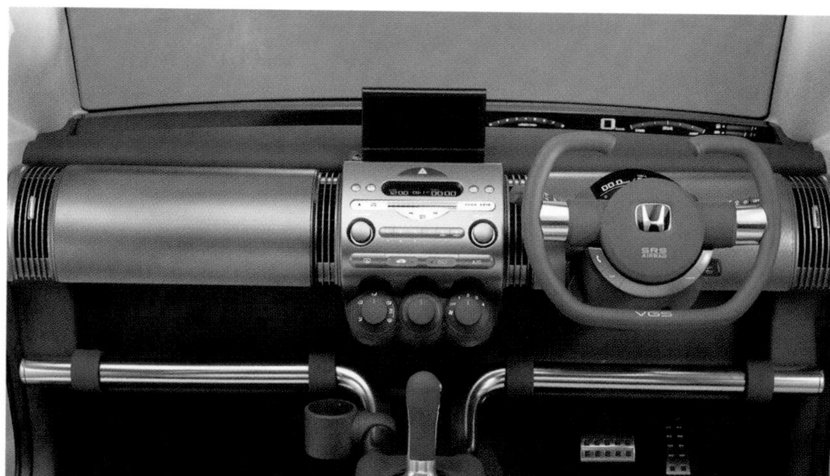

Car or motorcycle? With today's unpredictable traffic levels, who can tell what mode of transport is going to be appropriate? The radical Honda Bulldog offers both options.

Both BMW and Honda have car and motorcycle divisions, but there is rarely any synergy between them and their car-making counterparts. Some Japan-market versions of the 1981 City/Jazz small car did come with a fold-up Honda motorcycle as an option. The Bulldog, however, actually incorporates the cycle into its pugnacious design concept.

The Bulldog's flexible personal transportation is based on its ability to carry and recharge the e-DAX, a small, electrically powered, collapsible motorcycle that is stowed in a dedicated location behind the vertical rear tailgate. The car's SUV styling is supposed to summon up images of a metal bulldog, solidly built, with high rear haunches and a bulbous nose. Its surfaces are dramatic, with a high belt line, a curved windshield, and small side windows. One interesting feature is the colored spokes of the alloy wheels.

Despite the car's considerable height, Honda says the Bulldog should have excellent fuel economy and sprightly performance, thanks to a 1.5 ltr engine that uses its Integrated Motor Assist (IMA) system for high power and cleaner exhaust emissions. In the tubular interior, a data-card CCD camera and a 200 mm (8 ins) TV monitor system give real-time communication with e-DAX when it is out on the road.

This is an interesting Honda concept vehicle. The car/motorcycle combination could be an innovative solution to urban commuting that rivals would find tricky to emulate.

Honda CR-V

Design	Takahiro Hachigo
Engine	2.4 in-line 4
Power	119 kW (160 bhp) @ 6000 rpm
Torque	220 Nm (162 lb ft) @ 3600 rpm
Gearbox	5-speed manual/4-speed automatic
Installation	Front-engine/4-wheel drive
Front suspension	MacPherson strut
Rear suspension	Double wishbone
Brakes front/rear	Discs/discs
Front tires	205/70R15
Rear tires	205/70R15
Length	4536 mm (178.6 ins)
Width	1783 mm (70.2 ins)
Height	1682 mm (66.2 ins)
Wheelbase	2619 mm (103.1 ins)
Track front/rear	1534/1539 mm (60.4/60.6 ins)
Curb weight	1491 kg (3288 lb)
0–100 km/h (62 mph)	10 sec.
Top speed	177 km/h (110 mph)
Fuel consumption	9.4 ltr/100 km (24.9 mpg)
CO_2 emissions	216 g/km

Honda's new "Compact Recreational Vehicle" is the company's second-generation CR-V model. It is, in fact, a completely new car, including an all-new powertrain and interior. The first CR-V, introduced in 1997, established Honda as an extremely competent producer of "lifestyle" 4×4 off-road cars in its own right, the Japanese company having benefited little in this area from its fifteen-year relationship with Britain's Rover Group and its respected Land Rover division.

The new CR-V features stronger, more defined looks than the model it replaces, yet the gentle curves and radiuses have softened it, and made the overall shape easier on the eye. The exterior design echoes the Honda family image, as exemplified by models such as the Jazz and the Civic, scaled up to fit a larger application. These chunky proportions give it a utility persona. Around the car, the blackened lower edge below the sills, and the rugged wheels, accentuate its off-road potential.

Compared to the previous model, length and width are slightly increased, to give more luggage capacity and rear legroom, while the raised floor and 1.68 m (52 ft) height make the CR-V extremely easy to get into. Honda has also improved the two-way tailgate. It still has a separate, upper glass hatch, so light luggage can easily be placed in the back of the car, but this now forms part of the lower, side-hinged tailgate. Thus the glass hatch no longer needs to be opened first to open the tailgate.

The new CR-V is not exactly a design classic, but, just as in the case of the previous model, the key to its success is its user-versatility. And that seems only to have been bettered in this second incarnation.

CAW 225

Honda Dualnote

Design	Kobayashi Masahide and Suzuki Takahiro
Engine	3.5 V6
Power	294 kW (394 bhp)
Installation	Mid-engine/rear-wheel drive
Brakes front/rear	Discs/discs
Front tires	225/45ZR18
Rear tires	225/45ZR19
Length	4385 mm (172.6 ins)
Width	1725 mm (67.9 ins)
Height	1385 mm (54.5 ins)
Wheelbase	2570 mm (101.2 ins)
Fuel consumption	5.6 ltr/100 km (41.9 mpg)

Dualnote is an off-the-wall concept unveiled by Honda at the Tokyo Motor Show in October 2001. It is a four-seat, mid-engine, rear-drive sports car, intended for driving enjoyment, but also offering excellent environmental credentials.

Its design, which has more than a hint of Honda's acclaimed mid-engine sports car, the NS-X, focusses on optimized aerodynamics. At the front, its low nose incorporates a prominent spoiler for downforce, made possible by the absence of a front-mounted engine. A wedge-shaped conduit running through the doors helps direct air through intakes above the rear-wheel arches, which keeps Honda's 3.5 ltr V6 i-VTEC gasoline-electric hybrid powerplant—key to the Dualnote concept—cooled. An additional electric drive motor for the front wheels helps deliver some 300 kW (400 bhp) of power, while also achieving fuel economy of 5.6 ltr/100 km (41.9 mpg).

In the interior, the instrument panel is split into three displays, built around interactive voice-recognition technology, jointly developed by Honda and the Nippon Telegraph & Telephone Corporation (NTT). They comprise a three-dimensional display for speed, a center display for night vision, and an information monitor for email, Internet, and navigation.

Although designed to seat four people, a roadgoing Dualnote would necessarily offer compromised rear-passenger comfort, and limited luggage space, owing to the space its rear-mounted engine takes up.

While suggesting a type of fuel-efficient, four-seat Japanese Ferrari, the Dualnote has nothing like the style it needs to achieve that, no matter how miserly its fuel thirst. A market for a hybrid supercar, however, may well exist among wealthy sybarites with a social conscience. After all, Honda—along with Toyota—is the first car maker in the world to offer a hybrid production car.

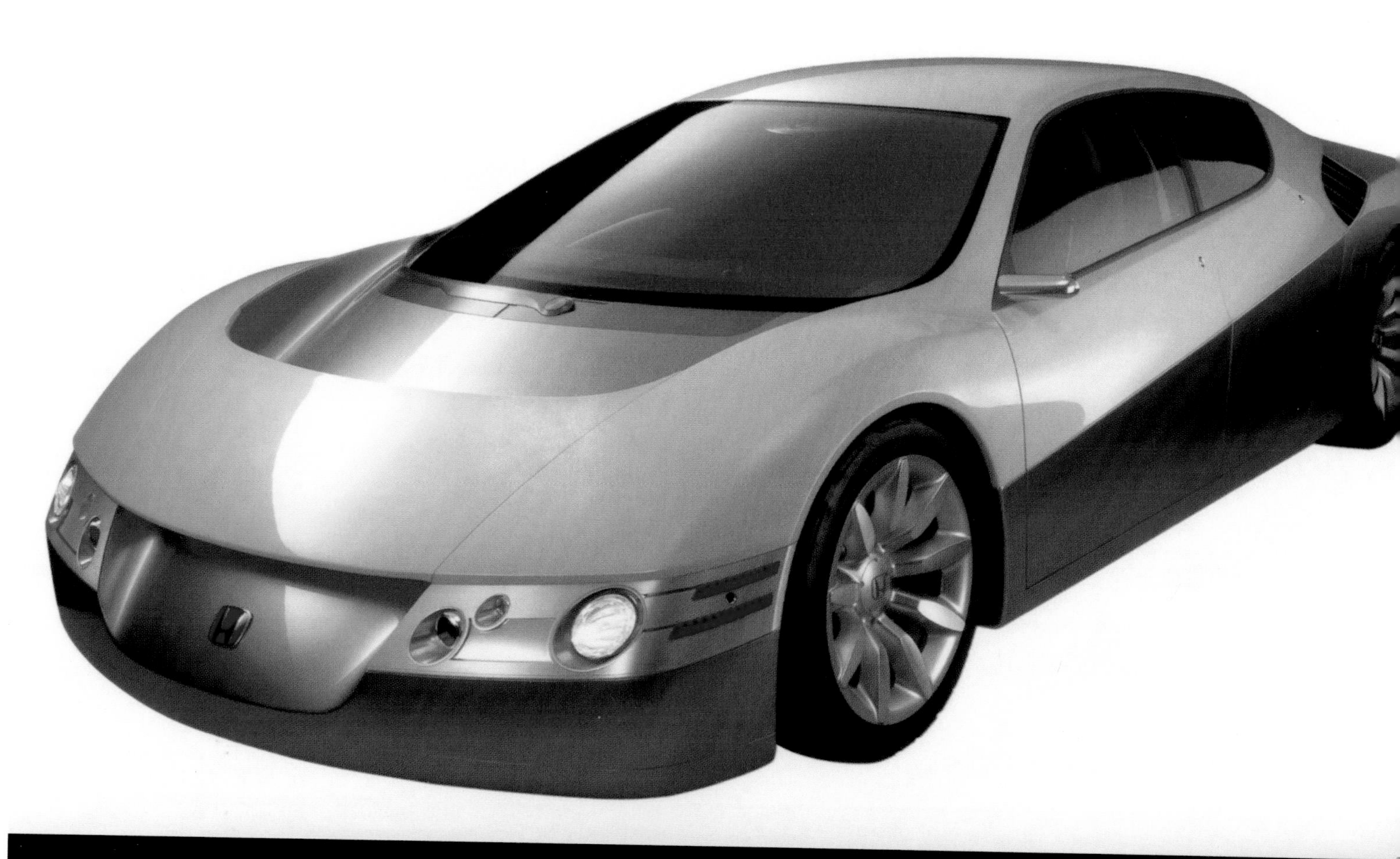

Honda Pilot

Design	Frank Paluch
Engine	3.5 V6
Gearbox	5-speed automatic
Installation	Front-engine/4-wheel drive
Front suspension	Independent strut-type, with "L" arm
Rear suspension	Independent multi-link, with trailing arm
Brakes front/rear	Discs/discs, ABS, EBD

North Americans show no sign of ending their love affair with big 4×4s, so Honda has created a new model, the Pilot, to pitch at best-sellers such as the Ford Explorer. Developed by Honda under the theme of "the ultimate family adventure vehicle," it has an exterior with classic SUV proportions: a strong stance, upright roof pillars, and a large "greenhouse" with panoramic views for all occupants. Rugged bumper molding details try to boost its aggressive look. At the same time, however, the overall shape and details are highly conservative—a contrast to visually exciting concepts such as the Acura RD-X (see p. 20). This reflects the huge financial risk of introducing a new model.

As befits a vehicle intended as a family carrier, the eight-seater Pilot was designed from the inside out, with three rows of seats, two of which fold away to boost cargo-carrying capacity. Both those rows (the second and third) also split-fold to allow a long load, such as a bicycle, to be squeezed in alongside a couple of passengers. In fact, thanks to a design that pushes its wheels nearer to the corners of the car, plus a flat cargo floor, the Pilot offers the largest, and most versatile, passenger- and luggage-carrying capability in its class.

The instrument panel has a sporty, three-gauge, center cluster inspired by a precision chronograph design. The steering wheel is decorated with silver trim, and carries buttons to control the stereo and cruise control. By moving the gear stick from a conventional, floor-mounted position to the steering column, Honda has cleared the center console area, making for a roomy, "walk-through" front cabin.

Developed and designed by Honda R & D Americas, the Pilot went into production in spring 2002. European sales, however, are unlikely.

Ricky Hsu 99

Honda Unibox

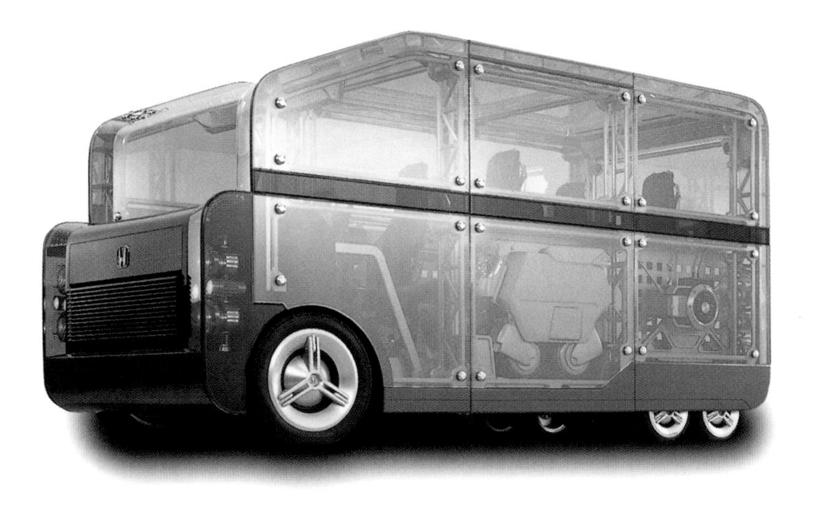

This concept "car", the Unibox, will surely be relegated to the history books of Honda design, despite the fact that it is both dramatic and, quite possibly, a real taste of the distant future. It is, in essence, an architectural dream from Honda's design department, a social and enjoyable space created in a futuristic glass box to inject some fun into life.

The Unibox is, as its name implies, a one-box design, with a strong architectural influence. Its tubular, aluminum truss frame is clad with flat modular panels, using deliberately visible fixture points.

The front is bus-like, with a large, rectangular grille, and a vertical row of colored lamps up each side. It is unapologetically rectangular, with similarly proportioned, rectilinear forms echoing throughout. The four small, partly-covered, rear wheels permit a low, flat floor, and so give maximum seating space.

Although the interior is expansive, and very well lit, the overall impression—from the outside—is of a cluttered, if functional, space. The extent of the clear panels helps to achieve Honda's objective, that the interior should be a relaxing place for passengers to sit back, chat, and drink coffee. We must wait and see whether Honda ever claims to have learned anything from this concept that can be implemented in its mainstream models.

Hummer H2

Design	Clay Dean and Terry Henline
Engine	6.0 V8
Power	242 kW (325 bhp)
Torque	502 Nm (370 lb ft)
Gearbox	5-speed automatic
Installation	Front-engine longitudinal/all-wheel drive
Front suspension	Torsion bar
Rear suspension	Live axle 5-link
Front tires	315/70R17
Rear tires	315/70R17
Length	4782 mm (188.3 ins)
Width	2469 mm (97.2 ins)
Height	1922 mm (75.7 ins)
Wheelbase	3114 mm (122.6 ins)
Curb weight	2597 kg (4525 lb)

Since the Gulf War dominated American TV screens in 1990, a certain type of customer has wanted a Hummer. A civilian version of the giant military off-roader has indeed been available since 1992, but its elephantine girth, vast weight, and tardy performance have limited sales to around one thousand fanatics a year. The Hummer H2, however, looks set to bring the no-nonsense, four-wheel-drive brand to a far wider audience when it goes on sale within the next two years. This is thanks to General Motors, which bought the Hummer marketing rights in 2000, and has spent the intervening time creating the H2, which was launched at the LA Auto Show in production form.

As a no-nonsense truck, the H2 has the unmistakably broad stance, and long wheelbase, characteristic of the original Hummer. Polished metal graces the vertical grille, and other functional exterior parts, for rugged looks and soldier-like durability. The exterior uses mainly rectangular forms and planar surfaces, the Hummer's uncompromising appearance softened slightly with small radiuses. Driving the H2 should give a heightened sense of security and protection, thanks to its strong underbody and upper structure.

Short overhangs, and huge ground clearance, allow large approach and departure angles, very useful when on trails, while massive towing hooks, and sturdy receiver hitches, are fitted to both ends of the vehicle. An optional winch attaches easily to either the front or the rear bumper, allowing the H2 to assist in many emergency situations.

When the bulkhead is in its upright position, the H2 can carry up to five passengers, in leather-covered seats that no GI has ever experienced. However, a roof-mounted, 360-degree, infrared, night-vision camera, and a GPS navigation system, emphasize the notion of this being a daily reconnaissance vehicle.

An SUT (sports utility truck) version, with an open back, was shown in concept form at the New York show in 2001. This aims to be an extremely tough model when it reaches the market, a vehicle that looks the part, and—crucially to the Hummer brand—is both functional and versatile. General Motors would be wasting its time if a macho-looking Hummer didn't also prove to be an extremely masculine machine in action.

Hyundai Clix

Design	Hyundai's Frankfurt design center
Engine	2.2 in-line 4
Power	225 kW (302 bhp)
Gearbox	6-speed sequential
Installation	Front-engine/4-wheel drive
Front tires	255/45R19
Rear tires	255/45R19
Length	3959 mm (155.9 ins)
Width	1738 mm (68.4 ins)
Height	1300 mm (51.2 ins)
Wheelbase	2450 mm (96.5 ins)

There has long been something of a chasm between Hyundai's constant procession of glittering concept cars and the mostly utilitarian fare on offer in its showrooms. Hyundai has christened the Clix a FAV—a fun activity vehicle—and, sure enough, there is no production Hyundai like it. Interestingly, however, this sporty two-plus-two originates not from Seoul, but from Hyundai's European design center in Frankfurt. Like the Saab 9X (see p. 222), it's a true "crossover" vehicle: the Clix can be transformed into a coupe, a pickup, a sports utility, or a convertible.

The Clix has a bullish design. Its dramatic front incorporates a huge, triangular grille, and large fenders that merge with the bumper, and encapsulate the headlamps. This makes the doors look small, an effect exaggerated by its trapezoidal-shaped side glass. The rear of the car, however, is altogether flatter, with a tall trunk opening, and vertically arranged rear lamps set back within bold glass housings.

The Clix has a striking all-glass roof, constructed of four separate panels. A high-tech opening mechanism uses six electric motors to power its transformation from coupe to targa, then to convertible and, finally, to pickup, with the panels stored under the rear seats.

This interesting concept lacks the design sophistication of Saab's 9X. Hyundai, however, launches new cars every few months, while new Saabs tend to come along at the rate of one every five years. Perhaps the first FAV you can actually own will indeed come from South Korea.

BORG
BORG
KIA

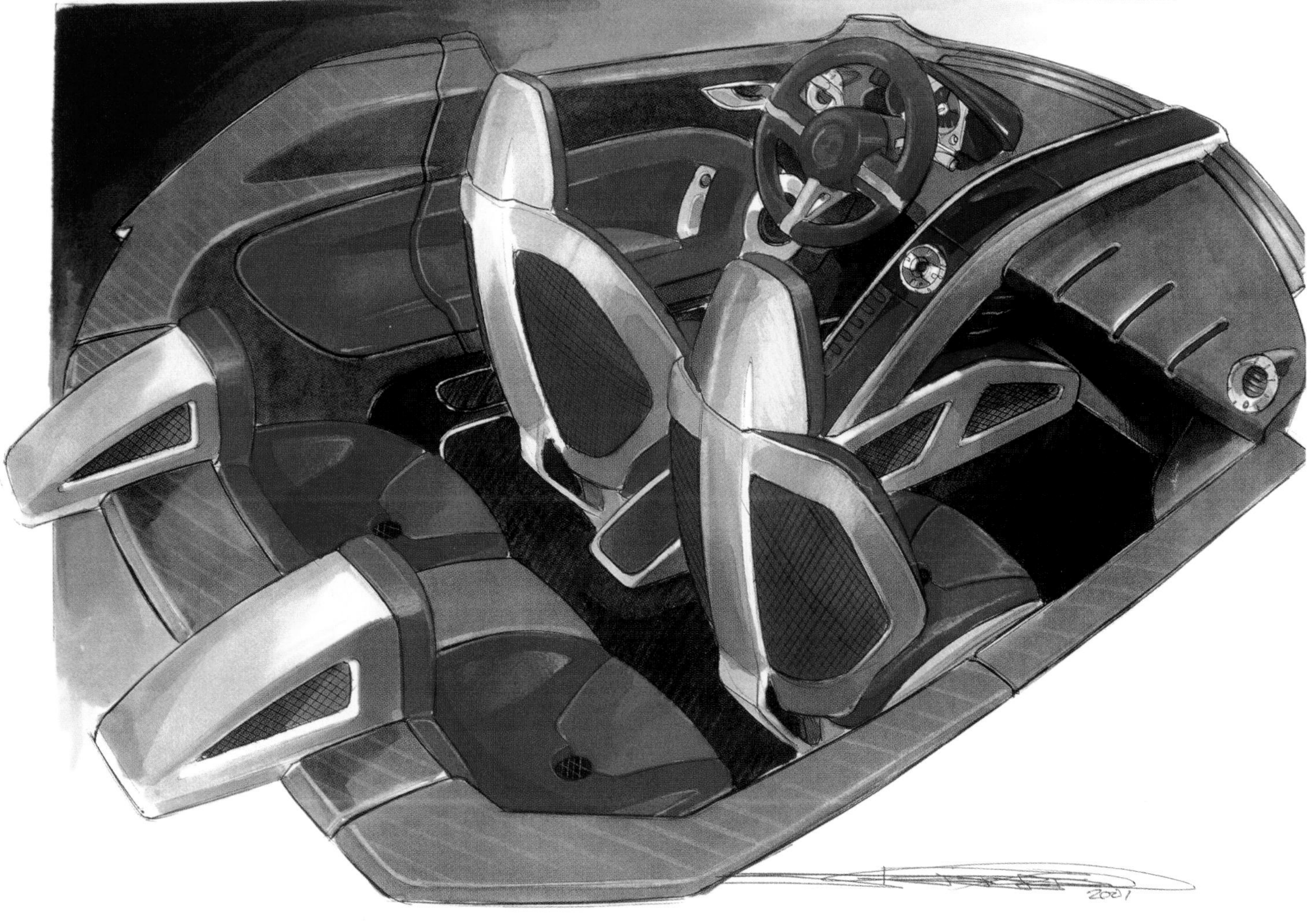
2001

Hyundai Getz

Design	Namyang Design Center, Seoul
Engine	1.6 in-line 4 (1.1 and 1.3 in-line 4, and 1.5 in-line 4 turbo-diesel, also offered)
Power	78 kW (105 bhp) @ 5800 rpm
Torque	143 Nm (105 lb ft) @ 3000 rpm
Gearbox	5-speed manual
Installation	Front-engine/front-wheel drive
Front suspension	MacPherson strut
Rear suspension	Torsion beam
Brakes front/rear	Discs/drums
Front tires	185/55R15
Rear tires	185/55R15
Length	3800 mm (149.6 ins)
Width	1600 mm (63 ins)
Height	1485 mm (58.5 ins)
Wheelbase	2450 mm (96.5 ins)
Track front/rear	1450/1440 mm (57.1/56.7 ins)
0–100 km/h (62 mph)	11.2 sec.
Top speed	182 km/h (113 mph)

The Getz is designed with a strong European flavor, because Hyundai expects 80% of its production to be exported to Western Europe, where the crucial B-car sector counts for 30% of sales. Maybe appropriately, it was unveiled at the Geneva International Motor Show in 2002.

The Getz platform is taken from Hyundai's Clix sports concept car (see p. 92) but it is a far more mainstream car, and is planned for high-volume production. There are three- and five-door versions. Its wheels are positioned close to the corners, and the high roof line banishes the claustrophobic feel of some small cars. Versatility is key to hatchback cars. The Getz has plenty of it, with a 60/40 split rear seat that also reclines and slides (with split double-folding cushion and back) for dozens of configurations.

Minimum safety standards ensure that the Getz is well equipped. The front seats are fitted with pre-tensioners, and height-adjustable seat belts. The central rear-seat space is fitted with a three-point seat belt, and a height-adjustable headrest of its own. Front occupants are protected by front and side airbags. Anti-lock brakes, with electronic brake-force distribution, are included on all models.

Neat if unremarkable, the Getz does not quite have the obvious desirability of, say, an Italian, German, or French small car, but Hyundais are usually competitively priced, so expect it to have lots of fans.

MTK HY 57

H HMCTB

Hyundai Tiburon Coupé

Design	Namyang Design Center, Seoul
Engine	2.7 V6 (1.6 and 2.0 in-line 4 also offered)
Power	122 kW (164 bhp) @ 6000 rpm
Torque	244 Nm (180 lb ft) @ 4000 rpm
Gearbox	Automatic
Installation	Front-engine/front-wheel drive
Front suspension	MacPherson strut, coil springs
Rear suspension	Multi-link, coil springs
Brakes front/rear	Discs/discs
Front tires	215/45ZR17
Rear tires	215/45ZR17
Length	4395 mm (173 ins)
Width	1760 mm (69.3 ins)
Height	1330 mm (52.4 ins)
Wheelbase	2530 mm (99.6 ins)
Track front/rear	1491/1491 mm (58.7/58.7 ins)
Curb weight	1333 kg (2939 lb)
0–100 km/h (62 mph)	8.3 sec.
Top speed	219 km/h (136 mph)
Fuel consumption	10.4 ltr/100 km (27.2 mpg)
CO_2 emissions	274 g/km

The new Tiburon Coupé was styled by Hyundai's Namyang design team, based in Seoul. Longer, wider, and taller than the previous model, it continues Hyundai's coupe lineage, established in 1989 with the Scoupe. South Korea's biggest brand has come a long way in the intervening thirteen years.

It doesn't take a detective, however, to work out where the car's sporty lines come from. Its similarity to the Ferrari 456GT is clear, particularly at the side. The 456GT is a modern classic, of course, but the visual relationship does nothing to dispel the nagging suspicion that Korean manufacturers tend to mimic rather than innovate.

Compared to the outgoing Tiburon, the styling is cleaner, more upmarket, and less aggressive. The distinctively creased fender bulges, rippled hood, and four separate headlamps are abandoned for a more sophisticated, patrician design: the new Tiburon has a conventional, V-shaped hood, and four headlamps mounted in pairs behind plastic covers. The tail is truncated, and the trunk lid has a small spoiler lip. Air vents sit just behind the front wheels, and powerfully sculpted creases run along the sides of the car to the rear fenders—all very Ferrari.

The Tiburon, however, is let down by one distinctly un-Italian touch: its wheels. They are too small for the high belt line and powerful haunches. The front three-quarter view is uncomfortable, too. It is simply not sharp enough.

Standard equipment levels are extremely high, a given with Hyundais. All models include driver, passenger, and side airbags, anti-lock brakes, a six-speaker radio/CD player, and air conditioning. Top models also have leather seats and cruise control.

Next to other big Asian car producers such as Honda and Toyota, Hyundai is second-league. But if the quality of the new Tiburon Coupé is good enough, the gap could close a little.

HYUNDAI

Infiniti FX45

Design	Shiro Nakamura
Engine	4.5 V8
Power	224 kW (300 bhp)
Torque	407 Nm (300 lb ft)
Gearbox	5-speed automatic
Installation	Front-engine/4-wheel drive
Front suspension	4-wheel independent, with multi-link rear
Rear suspension	4-wheel independent, with multi-link rear
Brakes front/rear	4-wheel vented discs, ABS, Brake Assist, EBD
Front tires	285/50R21
Rear tires	285/50R21
Length	4788 mm (188.5 ins)
Width	1948 mm (76.7 ins)
Height	1651 mm (65 ins)
Wheelbase	2850 mm (112.2 ins)
Track front/rear	1593/1651 mm (62.7/65 ins)

Luxury 4×4s are selling fast in North America. The Infiniti FX45 is an attempt by Nissan's luxury brand to seize a share of this market. Another "crossover" model, the FX45 matches flowing, coupe-style lines to the mud-plugging stance of a 4×4, and is the second concept to carry the name. In 2001 a less coherent design appeared at the Detroit auto show.

Powered by a V8 engine, the FX45 concept takes full advantage of its long wheelbase, wide stance, and 19 ins wheels and tires to establish a strong presence. Its cabin is moved rearward, set back like a sports car, with long doors, and wheels pushed out to the corners. A grille that drips chrome, a fluted hood, and narrow headlamps add a sporty feel, enhancing the Infiniti's attempt to attract buyers from rivals built by Cadillac, Lincoln, Land Rover, and Lexus.

Inside, the FX45 features brick-colored leather, and polished aluminum accents. An entertainment area is fitted with an overhead DVD display and game system, for rear-seat passengers. The conventional instrument panel is replaced with a large (243 mm/9.6 ins) LCD monitor.

The concept previews a production car for early 2003—although features such as the huge wheels will have to be toned down.

V8

Infiniti G35

Design	Mr Hasegawa
Engine	3.5 V6
Power	194 kW (260 bhp) @ 6000 rpm
Torque	353 Nm (260 lb ft) @ 4800 rpm
Gearbox	5-speed automatic
Installation	Front-engine/rear-wheel drive
Front suspension	Multi-link independent
Rear suspension	Multi-link independent
Brakes front/rear	Discs/discs, ABS, Brake Assist, EBD
Front tires	205/65R16
Rear tires	205/65R16
Length	4730 mm (186.2 ins)
Width	1750 mm (68.9 ins)
Height	1471 mm (57.9 ins)
Wheelbase	2850 mm (112.2 ins)
Track front/rear	1511/1514 mm (59.5/59.6 ins)
Curb weight	1528 kg (3369 lb)
0–100 km/h (62 mph)	6.2 sec.
Top speed	227 km/h (142 mph)
Fuel consumption	11.8 ltr/100 km (19.9 mpg)

By common auto-industry consent, the BMW 3 Series is one of the toughest of all acts to beat. It was the model that effectively created, and then monopolized, the highly profitable market for compact premium sports sedans. Mercedes-Benz, Audi, Jaguar, and Lexus have all fielded competitors, none of which has quite matched the BMW's abilities, or its magic. Now it is the turn of Infiniti, the upmarket arm of Nissan, to take a tilt at the German icon with its new G35.

The G35 has its origins in the 1999 Nissan XVL concept sports sedan shown at the Tokyo Motor Show in November 1999. Since that time Nissan's new controller, Renault, has brought its influence to bear on the external design of the model. It now offers a more aggressive front, with vertically stacked, clear-lensed headlamps, and a more graphic tail-light treatment.

Even so, the G35 does not have the kind of shape that would stand out in a crowd. Instead, its strength lies underneath, where it carries the mechanical elements of Nissan's fabled Skyline high-performance coupe. That counts for a lot among hard-driving enthusiasts, who have unstinting respect for the sophisticated Nissan sportster. Nevertheless, those judging purely on appearances will find the neat and uncontroversial Infiniti falling short on personality.

It is a similar story inside, where aluminum replaces wood to provide a sportier feel, but where design distinction is again lacking. Yet for many, especially the American clientele at whom the G35 is aimed, there is considerable appeal in the way this understated design belies a powerful mechanical specification that runs to an advanced 3.5 ltr V6 engine that gives no less than 194 kW (260 bhp), and, driving the rear wheels, a five-speed automatic gearbox, and the option of sports suspension. There is no way BMW could match this array for $27,000.

Irmscher Inspiro

Engine	3.0 V6
Power	165 kW (221 bhp) @ 5600 rpm
Gearbox	5-speed manual
Front tires	225/35R19
Rear tires	245/35R19
Curb weight	780 kg (1720 lb)
0–100 km/h (62 mph)	5.8 sec.
Top speed	240 km/h (149 mph)

Irmscher's Inspiro is a study for a purist's sports roadster. It aims to combine uncompromised driving pleasure with a classical roadster design.

This German company does not build complete cars. It does, however, sometimes unveil concept cars, such as this one, that help promote its wide range of performance and customizing accessories, perhaps most readily associated with Opel cars.

The two-seater Inspiro is defined by its extravagant design, meant to summon up halcyon images of classic sports cars from the 1950s and 1960s. Characteristics include the elongated hood, the arched, mudguard-like fenders, and the rear-biased, dragster-like seating position. The Inspiro's pugnaciously prominent radiator grille allows air to flow through the engine bay before leaving through two large, gill-like air ducts on the fenders.

The headlamps are mounted inboard, giving minimum overhang for the tire-hugging front fenders. Behind the snug cockpit, two protection bars, with matt-chrome finish, rise above the heads of the occupants, drawing the eye to raised body bulges on the rear trunk lid.

Driver and passenger alike wallow in a classic and functional ambience. The well-formed seats are leather-covered, and so are parts of the dash and leg space. Everything about the functional-looking controls shouts "drive me," especially the round instruments, embedded in a carbon-fiber pod, and easy to read behind the leather steering wheel.

Isuzu GBX

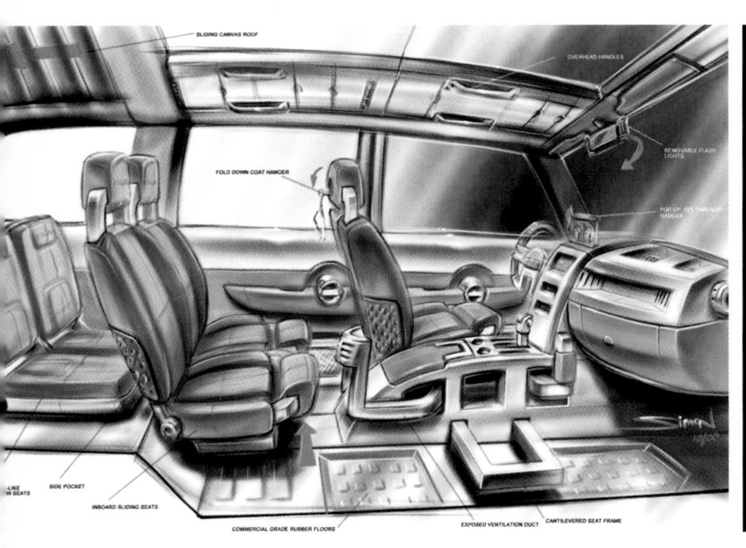

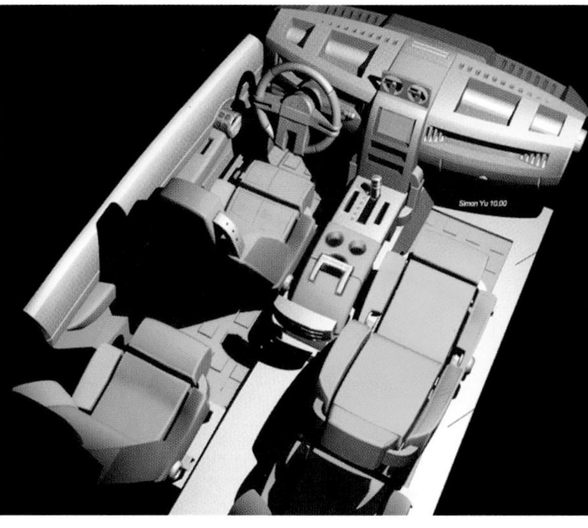

Design	Isuzu Motors America
Engine	3.0 V6 diesel turbo
Gearbox	4-speed automatic, with low range
Installation	Front-engine/4-wheel drive
Front suspension	Independent double wishbone, with torsion bar
Rear suspension	Multi-link, with coil springs
Brakes front/rear	Discs/discs
Length	4470 mm (176 ins)
Height	1891 mm (74.4 ins)
Wheelbase	2819 mm (111 ins)
Track front/rear	1515/1520 mm (59.6/59.8 ins)

Isuzu's GBX concept sports utility vehicle dates back some time: it was built by the Japanese company's California studio for the 2001 Detroit auto show. But, while its boxy appearance, and aggressive detailing, may already come across as somewhat dated, the concept contains many novel and practical ideas that could influence more mainstream future products in the Isuzu range.

The California-based designers sought to go back to SUV basics in configuring the GBX. Their idea was to create an "automotive backpack to transport your gear to wherever you want." The chosen inspiration was even older: the stagecoach, echoes of which are to be seen in the GBX's spoked wheels, its short overhangs, and its most distinctive feature, the bright metal, exoskeletal frame that surrounds the passenger cell. At the top, the frame becomes a roof rack. Deliberately tacked on to the base of this structure are rocker panels, also finished in bright metal, from which steps fold out, for ease of access to the cabin.

The GBX is, in fact, modular in construction: different nose and tail sections can be added to either end of the central passenger frame. The doors, following the current mode, are hinged at opposite ends, to provide unobstructed, pillarless access to the interior. Further novel ideas are to be found inside. The seats, for instance, are not mounted on the floor, but are cantilevered out from the central tunnel. This ingenious (but in all probability expensive) arrangement leaves the floor completely clear, so that long objects such as skis and snowboards can be easily carried. Even the trunk and rear door are carefully thought out. The door opens in two sections, one containing a tool kit for repairs to sports equipment, while the cargo area features storage units, shelving, and a canvas roof that can be extended when needed.

GBX
ISUZU

Isuzu Zen

The Isuzu Zen is a striking example of Japanese-influenced car design. The designers—working in Isuzu's European design studio, in Birmingham in the British West Midlands, under the leadership of design chief Geoffrey Gardner—were inspired by the Zen Buddhist principles of harmony, respect, purity, and tranquility.

The Zen's stylish front is clearly distinct from the van-like rear end, while its beautiful side windows are designed to represent the outline of a Japanese fan. The fan motif is repeated in the center console, and even on the tires. Sand-blasted window louvers running down the center of the roof echo a Japanese temple. The screen-type tailgate is inspired by the traditional Japanese home, horizontally split, and allowing for varying degrees of privacy and visibility.

Like a Transformer robot toy, the interior converts into a traditional Japanese tea room, complete with bamboo flooring, and Tatami woven mats. To achieve this, much of the conventional interior folds away, including the steering wheel and gear stick. The traditional seats, covered in royal-blue silk, disappear into the dash and floor, leaving a large open space intended to be ideal for contemplation or concentration.

The Zen's target market is made up of Tokyo high-flyers, who could turn their car into a stationary office during traffic jams or while parked near their next appointment. With its ample ground clearance, and Isuzu's four-wheel-drive system, the Zen is also suitable for a younger audience that wants to go off exploring. The incorporation of Japanese architecture and culture into the Zen makes it look fantastic, but would GM-controlled Isuzu ever develop a production version?

ISUZU
ISUZU
SKIN PANEL
SEAT IN POSITION ON DASHBOARD
INTEGRAL SEAT BELTS.
MOULDED SQUAB + BACKREST BONDED TO SKIN.
IAN 2001.
WOODEN LAMINATE SKIN.
CUSHIONS ON UNDERSIDE OF SEAT SKINS
SEAT BACK FOLDS
FAN GRAPHIC
SPRUNG PANEL/CUSHION IN FLOOR.

Italdesign Brera

Design	Giorgetto Giugiaro
Engine	4.0 V8
Power	298 kW (400 bhp) @ 7000 rpm
Gearbox	6-speed sequential
Installation	Front-engine/rear-wheel drive
Brakes front/rear	Discs/discs
Front tires	19 ins PAX run-flat system
Rear tires	19 ins PAX run-flat system
Length	4388 mm (172.8 ins)
Width	1894 mm (74.6 ins)
Height	1289 mm (50.7 ins)
Wheelbase	2595 mm (102.2 ins)
Track front/rear	1818/1884 mm (71.6/69.6 ins)

This mouthwatering Alfa was one of the stars of the Geneva International Motor Show in 2002, although Alfa Romeo had little to do with it. The car is the work of Giorgetto Giugiaro's Italdesign consultancy.

The Brera is a proposal for a Ferrari-V8-powered Alfa Romeo, that could easily be produced if Alfa wanted to move its focus upmarket. The brand already has a premium heritage in such production cars as the 1968 Montreal and the 1990 SZ, not to mention its glorious cars of the 1930s. The interesting 1996 Sbarro Issima concept was, like the Brera, a high-powered, front-engine sports car that could easily crown the Alfa Romeo range.

The Brera is powered by the V8 engine from the Ferrari 360 Modena. Using this thoroughbred power unit would help achieve market acceptance for such an expensive Alfa Romeo, just as the 1960s Fiat Dino, 1970s Lancia Stratos, and 1980s Lancia Thema 8.32 used Ferrari engines for credibility among influential enthusiasts.

This attractive concept uses many Alfa design cues seen today on the new 147. The Lamborghini-type scissor doors make an imposing statement about the Brera's performance personality, while the interior uses red leather and chrome to emphasize this sportiness. Italdesign's rendition of BMW's iDrive system is fitted, reducing the number of dash control interfaces, and design "clutter," in the cabin. The interior's most revolutionary aspect is its new sound system, designed by ATAO of Luxembourg. This eliminates the need for mechanical speakers; instead, air-molecule modulators reproduce sound without moving parts.

If produced by Alfa Romeo, the Brera would be in direct, and fierce, competition with the Maserati brand, also a Fiat-owned company. Still, it would be great to see the Brera as a production reality. If the positive press it received at Geneva is anything to go by, it would have a bright future.

Jaguar R-Coupé

Design	Ian Callum and Julian Thomson
Engine	V8
Installation	Front-engine/rear-wheel drive
Front tires	285/30R21
Rear tires	285/30R21
Length	4925 mm (193.9 ins)
Width	1890 mm (74.4 ins)
Height	1347 mm (53 ins)
Wheelbase	2909 mm (114.5 ins)

Ian Callum and Julian Thomson designed the voluptuous R-Coupé together, embodying a three-dimensional suggestion as to how Jaguar's design philosophy might develop. They incorporated the overall body shape and front grille characteristic of this make. The R-Coupé's proportions indeed suggest power and movement reminiscent of classic Jaguars. Equally fundamental to the character of a true Jaguar is the car's stance: the relationship between the road, the wheels, and the body. The R-Coupé's cabin, tucked down between its massive wheels, conveys an appropriate impression of power.

The surfaces of the car are complex, and elegantly understated. Its geometric forms, like its tubular fuselage, recall the E-type. Design detail is restrained, and calculated to deliver a sense of movement. The side-window trim subtly widens as it runs back, while the flowing belt line carries the eye along the length of the car—another Jaguar hallmark.

Modern furniture and interior design inspired both the styling and choice of materials for the R-Coupé's cabin. Wood and leather are used abundantly, to create a highly tactile interior. A broad sweep of Macassar ebony veneer around the cabin is offset by naturally treated leather. The blond Connolly hide on the seats contrasts with the deep-brown saddle leather used elsewhere, including on the floor. This rich detail throughout the cabin is, according to Jaguar, inspired by jewelry, watches, and luxury luggage.

The R-Coupé is not intended for production, but it embodies themes and ideas that may find their way into future models. It is not a typical Jaguar; they have tended to be either four-door sedans or two-seat sports cars. The rare, mid-1970s XJC is the nearest thing to an R-Coupé antecedent.

Jeep Compass

Design	Michael Castiglione
Engine	3.7 V6
Power	157 kW (210 bhp)
Torque	319 Nm (235 lb ft)
Gearbox	4-speed automatic
Installation	Front-engine/all-wheel drive
Front suspension	Independent, with upper and lower A-arms, coil springs
Rear suspension	Live axle, with trailing upper A-arm, dual trailing lower arms, coil springs
Brakes front/rear	Discs/discs
Front tires	235/55R20
Rear tires	235/55R20
Length	4150 mm (163.4 ins)
Width	1834 mm (72.2 ins)
Height	1636 mm (64.4 ins)
Wheelbase	2649 mm (104.3 ins)
Track front/rear	1544/1544 mm (60.8/60.8 ins)
Curb weight	1406 kg (3100 lb)
0–100 km/h (62 mph)	9 sec.
Top speed	176 km/h (109.3 mph)

Jeep needs to appeal to younger customers—a familiar motivation in the US—and the Compass is a study aimed at winning over this market.

A combination of coupe and 4×4, another "crossover" vehicle, the Compass has soft, rounded styling, and a sporty stance. These represent an attempt at capturing the imagery of cars often seen in computer games, which Jeep believes to be a strong influence on its target market of "millennials": car buyers under twenty-four. Jeep has released a strong line of concepts over the past eight years. The Compass displays their influence, including that of the 1998 Jeepster concept, itself a blueprint for the new Jeep Cherokee.

Key Jeep design themes, such as the seven-bar grille, round headlamps, and bulging front fenders, define the look of the Compass. New design language includes fender-length running lights, stainless-steel brush guards and sill panels, inscribed with the Jeep name, and a roof featuring a diamond pattern inspired by the steel plate used in factories. The "Force Green" exterior paint is reminiscent of a military color scheme, and is steeped in Jeep authenticity.

With Jeep working hard to improve the standard of its production-vehicle interiors, studies such as the Compass, with its cabin inspired by that of a fighter aircraft, help. Technical dials and gauges, including—of course—a compass, abound. But the styling is as uncluttered and functional as possible. Slot-machine-style rotating controls operate all primary functions, while the gauges evoke traditional watch faces—a feature explored in the Jeep Varsity concept of 2000—and the air vents have a classic, aeronautical look. The four bucket seats are trimmed in green leather and Goretex. The interior features a molded rubber floor, holding true to Jeep heritage.

Jeep

Kia Sorento

Design	Byung-jun Min and Nak-Chul Ki
Engine	3.5 V6 (2.4 in-line 4 and 2.5 in-line 4 diesel also offered)
Power	143 kW (192 bhp) @ 5500 rpm
Torque	300 Nm (221 lb ft) @ 3000 rpm
Gearbox	4-speed automatic
Installation	Front-engine/rear- or 4-wheel drive
Front suspension	MacPherson strut, double wishbone
Rear suspension	Multi-link
Brakes front/rear	Discs/discs
Front tires	245/70R16
Rear tires	245/70R16
Length	4567 mm (179.8 ins)
Width	1960 mm (77.2 ins)
Height	1730 mm (68.1 ins)
Wheelbase	2710 mm (106.7 ins)
Track front/rear	1580/1580 mm (62.2/62.2 ins)
Curb weight	1929 kg (4253 lb)

Kia announced its 2003 Sorento at the Chicago auto show in 2002. It is a larger, more powerful SUV than the Korean maker's established, and reasonably popular, Sportage.

A rugged, authentic SUV, with a generous hint of Mercedes-Benz M-Class styling, the Sorento features straightforward, ladder-frame chassis construction, with double-wishbone front suspension, and multi-link rear suspension. A long wheelbase and wide track should provide a comfortable ride and good stability, while short front and rear body overhangs allow the Sorento to tackle steep terrain.

Both two- and four-wheel-drive configurations are available. In the two-wheel-drive version, power is directed to the rear wheels, while the 4×4 offerings include two different drive systems, one part-time and one on demand. The part-time four-wheel-drive system engages with the turn of a knob for convenient, "shift-on-the-fly" driving. Additionally, the Sorento offers an Eaton carbon, limited-slip differential that quietly allocates up to 50% of the torque between the rear wheels.

Inside, Kia's Sorento incorporates plenty of storage space, an underbody-mounted, full-size spare tire, a flat cargo floor, a split-opening rear hatch, an under-seat storage box, and no less than eight cup-holders. It also seats five passengers, with generous headroom and legroom.

In addition to hauling cargo, the Sorento can tow up to 1590 kg (3506 lb), and comes pre-wired for trailer lights. A load-leveling system is available to keep the car level while towing or carrying heavy loads.

Mercedes styling with plenty of performance and equipment, aimed at the US market at a likely rock-bottom price: that just about sums it up.

KIA

Koenigsegg CC 8S

Engine	4.7 V8 supercharged
Power	488 kW (655 bhp) @ 6800 rpm
Torque	750 Nm (553 lb ft) @ 5000 rpm
Gearbox	6-speed manual
Installation	Mid-engine/rear-wheel drive
Front suspension	Double wishbone
Rear suspension	Double wishbone
Brakes front/rear	Discs/discs
Front tires	245/40ZR18
Rear tires	315/40ZR18
Length	4190 mm (165 ins)
Width	1990 mm (78.3 ins)
Height	1070 mm (42.1 ins)
Wheelbase	2660 mm (104.7 ins)
Curb weight	1170 kg (2580 lb)
0–100 km/h (62 mph)	3.5 sec.
Top speed	390 km/h (242 mph)
Fuel consumption	14 ltr/100 km (16.8 mpg)

The Koenigsegg CC 8S project was initiated in Sweden eight years ago. After the prototype was revealed at the 1997 Cannes Film Festival, the positive response it generated led directly to a decision to launch a production version.

The CC 8S is a two-seat mid-engine supercar that makes extensive use of Formula 1 technology. Its quad-cam V8 engine delivers 750 Nm (553 lb ft) of torque, and weighs only 240 kg (529 lb), achieved by using carbon fiber, titanium, and aerospace-standard aluminum. The body and chassis are made of autoclaved epoxy pre-impregnated carbon fiber, making the car extremely strong, stiff, and lightweight. It has twenty-one layers of carbon fiber, varying between unidirectional and woven material, with intermediate spacing of aluminum honeycomb, as well as integrated, tool-grade, solid-aluminum joints. The total weight of the monocoque is just 62 kg (137 lb).

The interior of the CC 8S is trimmed with leather, aluminum, and more carbon fiber. The design uses a dynamic flow of lines that originate in the console, and run symmetrically along the windshield. The focus of the interior is the circular main control unit, which has stainless-steel buttons, and a multicolor zodiac of lights and icons. The carbon-fiber chassis is left visible, to expose its high-tech construction.

The avant-garde doors are unique. A geared rotational pivot operates simultaneously with a parallel arm, in an outward arc. Balanced by gas struts, the doors lift easily to allow entry, their refined, "flying-punch" movement setting the stage for this mechanical drama.

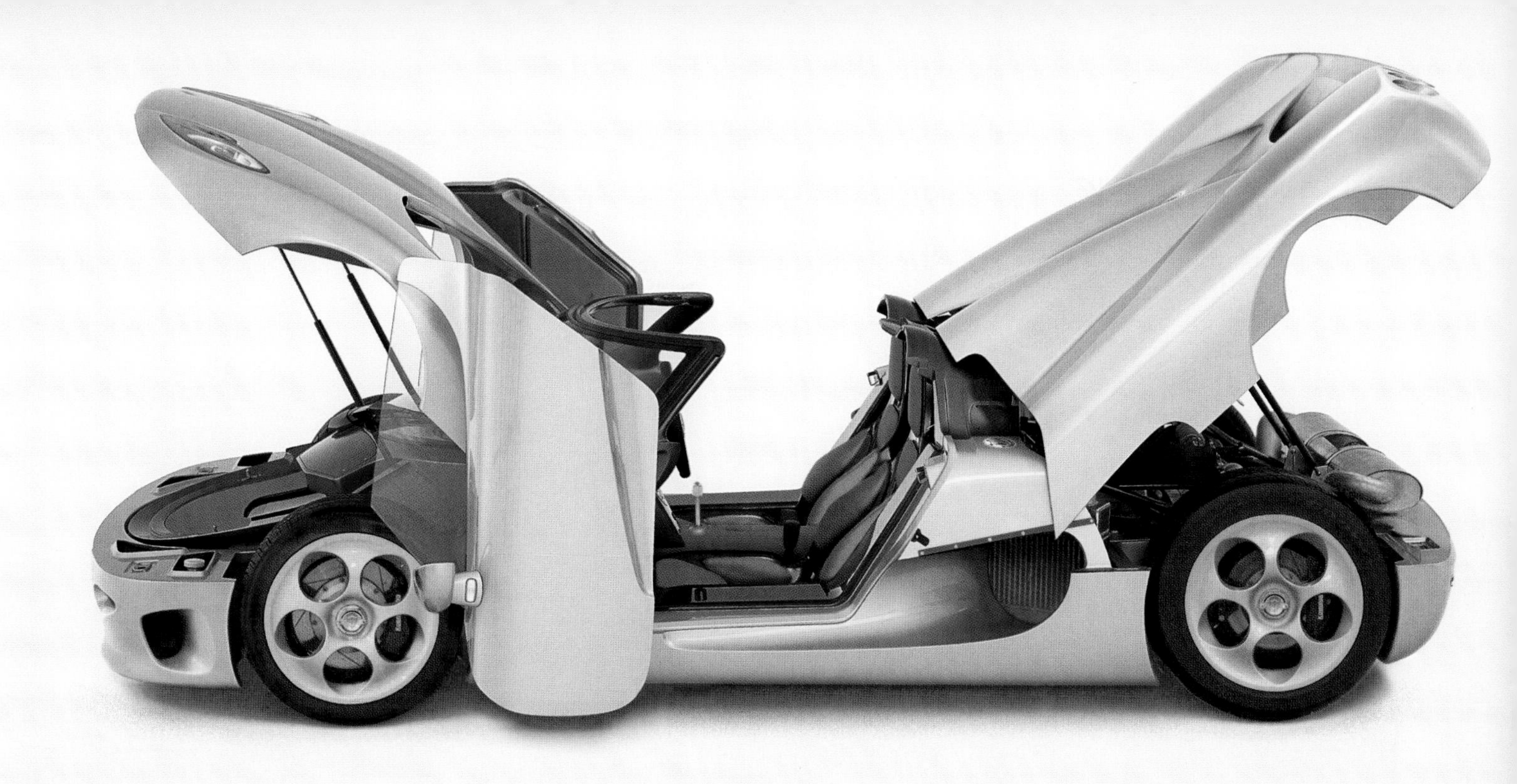

Koenigsegg CC

Koenigsegg
CC 8S

Lamborghini Murciélago

Design	Luc Donckerwolke
Engine	6.2 V12
Power	426 kW (572 bhp) @ 7500 rpm
Torque	650 Nm (479 lb ft) @ 5400 rpm
Gearbox	6-speed manual
Installation	Mid-engine/4-wheel drive
Front suspension	Independent, double wishbone
Rear suspension	Independent, double wishbone
Brakes front/rear	Discs/discs
Front tires	245/35ZR18
Rear tires	335/30ZR18
Length	4580 mm (180.3 ins)
Width	2045 mm (80.5 ins)
Height	1135 mm (44.7 ins)
Wheelbase	2665 mm (104.9 ins)
Track front/rear	1635/1695 mm (64.4/66.7 ins)
Curb weight	1650 kg (3638 lb)
0–100 km/h (62 mph)	4 sec.
Top speed	330 km/h (205 mph)

This is Lamborghini's successor to the Diablo, named Murciélago after a fighting bull whose life was spared—an honor bestowed only on beasts that show exceptional courage and spirit in the arena. The name harks back to the glory days of the classic Miura, also named after a bull.

According to the design brief, the car had to be exciting and unmistakable, while also safe and ergonomic. The Murciélago is a two-seat, two-door coupe based on a traditional Lamborghini layout: a mid-mounted V12 engine, with the transmission mounted in front of the engine. This layout has been employed by Lamborghini for over thirty years, in fact since the Miura made its bow in 1966, and gives optimum weight distribution. The addition of four-wheel drive in the Murciélago, like the Diablo, gives traction, braking, and handling advantages.

Central to the Murciélago's design are purity of line, muscularity, aeronautical influence, efficiency, and a "made to measure" quality. The result is that the new car has a simpler form than previous models. A single arc extends from the front to the rear, emphasizing the overall wedge shape of the car. The external bodywork panels are made of carbon fiber (for lightness and strength), except for the steel roof and the panels for the upward-opening doors—a Lamborghini hallmark since the first Countach in 1972.

The design of the Murciélago's aerodynamic surfaces originated in aeronautical engineering. Function therefore dictated form. The rear of the car features a variable air-flow cooling system (VACS) which can alter airflow to the engine, depending on driving conditions. The angle of the rear spoiler can also be altered, to optimize aerodynamic downforce.

The Murciélago is probably the most comfortable two-seater Lamborghini has ever produced, with a trip computer, leather upholstery, CD changer, and navigation system. Interior noise has also been suppressed by masses of sound-deadening insulation. This powerful, aggressive-looking machine is instantly recognizable as one of the most exclusive sports cars in the world.

Lancia Phedra

Engine	3.0 V6 (2.2 in-line 4 turbo-diesel also offered)
Power	150 kW (204 bhp)
Length	4750 mm (187 ins)
Width	1860 mm (73.2 ins)
Height	1750 mm (68.9 ins)

The Phedra replaces Lancia's Zeta full-size minivan. It arrives at a difficult time for the troubled Lancia brand, sales of which have declined 13.7% in Western Europe during 2001.

The Zeta sold just 1764 units in 2001, so the new Phedra must perform much better to contribute to Lancia's survival. The Phedra, created as part of the well-established joint venture with Peugeot, is bigger than the previous model. Taut, clean lines, and architectural shapes with a strong Italian flavor, make the Lancia Phedra immediately recognizable, and give it pronounced personality.

The front end is sculpted, and dominated by distinctive Lancia motifs. The bumpers are fully integrated with the car's surface, and the large, upright grille conveys a sense of superiority rather than aggression. Light clusters are gem-like and clean-shaped, in line with Lancia's brand philosophy.

The long sides of the body display lines that run crisply toward the rear, and shapes marked by slightly angled corners. The two door handles stand side by side (a hint of chrome is reflected in the side strips and on the grille) as a reference to great Lancias of the past, such as the Aurelia sedan of the 1950s. Another distinctive feature is door mirrors that incorporate the side turn signals—a Mercedes-Benz idea.

The Phedra's passenger space is lounge-like, an impression reinforced by extensive use of Alcantara, together with mahogany-pattern inserts, and burnished chrome features.

LANCIA PHEDRA
PHEDRA

LANCIA PHEDRA

Lexus GX470

Design	Yasushi Nakamura and Norihito Iwao
Engine	4.7 V8
Power	175 kW (235 bhp)
Torque	434 Nm (320 lb ft)
Gearbox	5-speed automatic
Installation	Front-engine/4-wheel drive
Rear suspension	Air suspension, with self-leveling and standard Adaptive Variable Suspension damping
Brakes front/rear	ABS, Brake Assist, DAC, EBD, traction control, and VSC
Front tires	265/65R17
Rear tires	265/65R17
Length	4780 mm (188.2 ins)
Width	1880 mm (74 ins)
Wheelbase	2789 mm (109.8 ins)
Curb weight	2087 kg (4602 lb)

Lexus has scored a huge hit in North America with the RX300, a compact 4×4 that drives, and looks like, a car. At the top of its range in America it also sells the LX470, a re-emblemed version of the Toyota Land Cruiser 500. Between the two is a market niche for a vehicle that seats eight like the LX470, but is more compact and therefore sharper to drive. Enter the GX470.

Featuring the same permanent four-wheel-drive system as the premium LX470, as well as its 4.7 ltr V8 engine, and rear adaptive height controls, the GX470 is clearly designed with the other Lexus 4×4s in mind with contemporary styling, and aggressive features such as a bold grille, and striking bumper air intakes. Plastic cladding—panels added over the bodywork to change the car's styling—protect the body, while emphasizing the rugged nature of the GX.

Lexus interiors tend be well executed but bland. The GX follows this trend, mixing gray leather with wood veneer to give a well-established, luxury look for the US market. Luxury features typical in this class feature in the cabin, including leather trim, a premium audio system, and electric seats, and its steering wheel retains a memory of the driver's preferred positions. Newer features include an optional, factory-installed DVD entertainment system that descends from the ceiling for the rear passengers, and an overhead console with integrated "HomeLink" transmitter.

These features, matched to Lexus's strong image in North America, will provide stiff competition for home-grown and import 4×4s, although the GX 470's styling still lacks presence.

GX470

Lexus Movie

Design	Steven Spielberg, Harald Belker, and Calty (Toyota/Lexus design studio)
Engine	Smart recharging electric engine
Power	500 kW (670 bhp)
Front suspension	Titanium double wishbone
Rear suspension	Titanium double wishbone
Brakes front/rear	Discs/discs
Front tires	285/30R22
Rear tires	285/30R22
Length	3708 mm (146 ins)
Width	2083 mm (82 ins)
Wheelbase	2692 mm (106 ins)
Curb weight	1043 kg (2300 lb)

The name "Movie" might seem pretty strange for a concept car, especially one as outrageous as this Lexus-emblemed design. But the label, and also the extreme nature of the design, begin to make much more sense when you realize that this self-styled concept car of the future was created specifically for *Minority Report*, a Steven Spielberg film released by Twentieth Century Fox in the summer of 2002.

Minority Report is set in the year 2054, in a future judicial system in which killers are arrested and convicted before they commit murder. Spielberg, himself a Lexus owner, needed vehicles that would convey a suitably futuristic impression. He approached the company to research and design a sports car specially for the film.

Fifty years into the future is well beyond the planning horizon of even the most far-sighted car company. Accordingly, Spielberg, Lexus's California-based design studio, and a team of futurists, including conceptual artist Harald Belker—whose film design credits include *Batman and Robin,* and *Armageddon*—met early in the development process to speculate on how automotive travel might have evolved by the middle of the century.

The Movie concept was the result. It represents a vision of a high-performance, two-seat, personal sports car of the year 2054. Freed from the restrictions of current regulations, and present-day drivetrain and chassis technology, the Movie's design is intriguing, and also ambiguous, in that it is not immediately clear which is the front and which the rear. Muscular design, with a cab-forward driving position, blends with racy-looking, exposed wheel sets, and a wide, but truncated, front that challenges familiar proportions and preconceptions.

In terms of brand awareness, the placement of this vehicle in such a high-profile movie is highly beneficial to Lexus. But, as many car manufacturers have found out to their cost, the link between the exciting designs of the future and the more humdrum machines of today can be difficult to establish.

Lincoln Continental

Design	Gerry McGovern
Engine	6.0 V12
Power	309 kW (414 bhp) @ 6000 rpm
Torque	560 Nm (413 lb ft) @ 5270 rpm
Gearbox	6-speed automatic
Installation	Front-engine/rear-wheel drive
Front suspension	Multi-link independent, with driver-selectable electronic damping
Rear suspension	Multi-link, with driver-selectable electronic damping
Brakes front/rear	Discs/discs
Front tires	275/45R22
Rear tires	295/40R22
Length	5444 mm (214.3 ins)
Width	1950 mm (76.8 ins)
Height	1500 mm (59.1 ins)
Wheelbase	3470 mm (136.6 ins)
Track front/rear	1675/1655 mm (66/65.2 ins)
Curb weight	1745 kg (3848 lb)

Lincoln, Ford's most prestigious US brand, and one that traditionally rivaled Cadillac for kudos, was generally felt to have lost its way in the 1980s and 1990s. The task faced by incoming chief designer Gerry McGovern was to build a new and distinctive identity for the fading brand, although Lincoln now sits outside Ford's Premier Automotive Group.

Two years in the making, McGovern's study for a new-era Lincoln Continental—the brand's largest and most famous model – can be taken as a strong statement of the direction Lincoln intends to adopt. The new Continental aims to highlight qualities such as elegance, simplicity, precision, and overall restraint, hallmarks of the Continental in its heydays of the 1940s, 1950s, and 1960s, and, Lincoln executives believe, still relevant today.

The exterior design is suitably understated. Large, gently curved surfaces give a sophisticated feel, echoing that of the 1960s Continental. The horizontal emphasis continues at the rear of the car, with an uncluttered rear end, and large LED lamps that split in the middle to allow positioning of a Lincoln star emblem. The four rectangular exhaust pipes, sunk into the rear valance panel, hint at the substantial power available from the Aston Martin-derived V12 engine.

Vast overall dimensions, and large center-opening doors, a feature of past Lincoln models, ensure easy access to the huge interior. The ceiling and seats (inspired by the McGovern-favored, design-classic Eames lounge chair) are covered in full-grain aniline leather. The carpeting is a close-sheared, midnight-blue sheepskin.

As an indicator of the Lincoln brand's direction, the Continental concept points clearly toward a golden age when the brand stood for elegance, dignity, and discreet, yet expansive, style. Few would dispute that this is the right way for Lincoln to go in the twenty-first century.

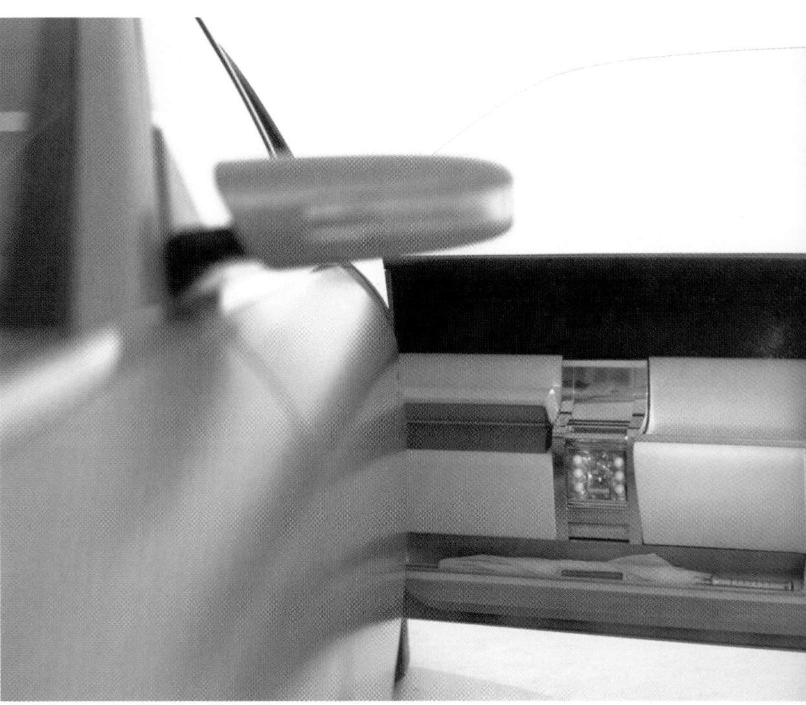

LINCOLN

Lincoln MK 9

Design	Gerry McGovern
Front tires	275/45R22
Rear tires	295/40R22
Length	5260 mm (207.1 ins)
Width	1950 mm (76.8 ins)
Height	1423 mm (56 ins)
Wheelbase	3095 mm (121.9 ins)

Gerry McGovern, the flamboyant Briton two years into his stint as design director of Lincoln, is used to redefining the direction of "problem" brands. In previous jobs, he was responsible for pointing both MG, with the MGF, and Land Rover, with its Freelander, toward the future.

With a fading, all-American image as a traditional old man's car, Lincoln has been experiencing a makeover under McGovern's unerring eye. And the all-new Lincoln MK 9 coupe concept is where it's heading.

A few of Lincoln's traditional trademarks survive. The MK 9's striking face, for instance, incorporates a new version of the prominent Lincoln grille, albeit flanked by twin Xenon headlamps. The chrome strip that defines its high-level belt line instantly draws the eye, and accentuates the 5.2 m (17.25 ft) length, while gloss-black paint clearly highlights the body surfaces. The massive 22 ins wheels are the dream of car designers, but would probably jeopardize ride and handling. They may be made smaller to keep Lincoln's serenity-craving customers happy.

With memories of tacky Lincoln interiors of the 1970s and 1980s still vivid in the minds of wary non-owners, McGovern has sought inspiration from a more admired American source—Charles Eames. His Lounge Chair, a mid-twentieth-century American classic, designed with comfort in mind, is given a nod by the MK 9's front seats, which are cantilevered off the center console to improve passenger foot space.

The interior and exterior are connected by body-colored seat shells with horizontal chrome finisher. A combination of dark-cherry and "Marlboro"-red leathers, with accents of polished metal, creates a luxuriously enticing environment. Dark-cherry leather is also used instead of carpeting, while the head-lining is white leather. A screen displaying navigation and telematics information dominates the center console. McGovern intends the etched-glass instruments to have a jewel-like quality—an aura he has aimed for before on many MGF details.

With the MK 9, Lincoln wants to maintain its American-luxury theme while creating an exciting car. Maybe this huge car has the right gangster-movie touch to do just that.

Coupe
MK9

Lincoln Navigator

Design	Gerry McGovern
Engine	5.4 V8
Power	224 kW (300 bhp) @ 5000 rpm
Torque	481 Nm (355 lb ft) @ 2750 rpm
Gearbox	4-speed automatic
Installation	Front-engine/4- or rear-wheel drive
Front suspension	Independent control arms, coil springs
Rear suspension	Independent control arms, coil springs
Brakes front/rear	Discs/discs, ABS, EBD
Front tires	255/70R18
Rear tires	255/70R18
Length	5232 mm (206 ins)
Width	2037 mm (80.2 ins)
Height	1976 mm (77.8 ins)
Wheelbase	3018 mm (118.8 ins)
Track front/rear	1701/1704 mm (67/67.1 ins)
Curb weight	2740 kg (6042 lb)
Fuel consumption	16.6 ltr/100 km (14.1 mpg)

Weighing almost three tons, seating eight people, and occupying 11 m² (13 sq yd) of road space, the Lincoln Navigator is a giant even among the already oversized species of American luxury sports utility vehicles. In its updated 2003 iteration it has grown still more sumptuous, more complex, and even better equipped.

The Navigator has an important position to defend. Long known as America's plushest, and most powerful, SUV, it is under threat from the new Cadillac Escalade, with no less than 224 kW (300 bhp). Lincoln, Ford's premium US brand, has responded with a whole host of innovations to spice up the familiar Navigator, which is derived from the Ford Expedition. Foremost among these are power-operated functions, to make life even more effortless for the Navigator captain and crew. The pedals, for instance, now adjust electrically, with a memory position function. There is no need to struggle to open the tailgate, or fold the rearmost row of seats; power takes care of these tiresome operations as well.

The running boards are also power-operated. In what Ford and Lincoln describe as an industry production first, these fold up flat against the body side to maintain smooth looks, then power out to ease access to the high cabin. Automatic lowering of the suspension helps too.

In terms of design, little has changed. The famous waterfall grille has become even more prominent. Inside the cabin, the symmetrical instrument panel, which claims to have been inspired by the 1961 Continental, is now trimmed with satin-finished nickel and American walnut. The vast front armchairs cool as well as heat. There is a navigation system in the front, and a DVD entertainment system for the two rear rows of seats.

These being the essentials of life in the gas-guzzling, heavy-luxury, US SUV market, the Navigator's 40% share of the premium business does not appear to be under serious threat quite yet.

NAVIGATOR

LINCOLN
NAVIGATOR

Maserati Coupé

Design	Giorgetto Giugiaro
Engine	4.2 V8
Power	287 kW (390 bhp) @ 7000 rpm
Torque	451 Nm (333 lb ft) @ 4500 rpm
Gearbox	6-speed manual/6-speed semi-automatic
Installation	Front-engine/rear-wheel drive
Front suspension	Double-link, with coil springs
Rear suspension	Double-link, with coil springs
Brakes front/rear	Discs/discs, ABS, EBD
Front tires	235/40Z18
Rear tires	265/35Z18
Length	4303 mm (169.4 ins)
Width	1822 mm (71.7 ins)
Height	1305 mm (51.4 ins)
Wheelbase	2440 mm (96.1 ins)
Track front/rear	1525/1538 mm (60/60.6 ins)
Curb weight	1730 kg (3815 lb)
0–100 km/h (62 mph)	4.9 sec.
Top speed	283 km/h (176 mph)
Fuel consumption	16.1 ltr/100 km (14.5 mpg)
CO_2 emissions	430 g/km

Even when it comes to exotic cars for the seriously rich, appearances can be deceptive. The new Maserati Coupé has a comfortingly familiar look to it, and even someone who knows their cars well would take it for the very pretty 3200GT, which marked Maserati's recent comeback to the high-class, grand-touring market.

Look closer, however, and there is an odd feel to a couple of details. The rear lamps, in particular, are no longer elegantly thin, boomerang-shaped strips that follow the shape of the body, but are now larger, and more conventionally placed. And the nose is almost imperceptibly different, too.

But few, if any, would suspect that this is an entirely different car, so successfully does the new coupe ape the shape and stance of the old 3200GT. The new car has a shorter wheelbase, a new and bigger 4.2 ltr engine, revised transmission choices, and the option of a "skyhook" adaptive-suspension system. Most of these changes were introduced on the Spyder convertible in 2001, so the new coupe completes the rationalization process, geared to the company's now very advanced manufacturing facilities. Those facilities provide very much greater capacity than Maserati has had in the past, capacity aimed squarely at the US, where the brand is relaunching in 2002 in perhaps the biggest expansion drive in its seventy-six-year history.

As in Europe and Japan, the Maserati Coupé and Spyder are positioned close to the very top of America's exclusive sports-car market, with only their sister brand Ferrari above them. The updated look, penned by Giorgetto Giugiaro of Italdesign, retains the curvaceous and muscular feel so characteristic of Italian exotic cars. The classic, jaw-like Maserati grille further emphasizes the brand's glorious heritage. With a nod to the leisure pursuits of US customers, the coupe's trunk swallows two golf bags, while the open Spyder is sure to be well received in Florida and California.

MO 390HP

MO 390HP

Maserati Spyder

Design	Giorgetto Giugiaro
Engine	4.2 V8
Power	287 kW (390 bhp) @ 7000 rpm
Torque	451 Nm (333 lb ft) @ 4500 rpm
Gearbox	6-speed manual/6-speed semi-automatic
Installation	Front-engine/rear-wheel drive
Front suspension	Double-link, with coil springs
Rear suspension	Double-link, with coil springs
Brakes front/rear	Discs/discs, ABS, EBD
Front tires	235/40Z18
Rear tires	265/35Z18
Length	4303 mm (169.4 ins)
Width	1822 mm (71.7 ins)
Height	1305 mm (51.4 ins)
Wheelbase	2440 mm (96.1 ins)
Track front/rear	1525/1538 mm (60/60.6 ins)
Curb weight	1730 kg (3815 lb)
0–100 km/h (62 mph)	4.9 sec.
Top speed	283 km/h (176 mph)
Fuel consumption	16.1 ltr/100 km (14.5 mpg)
CO_2 emissions	430 g/km

Maserati's new Spyder has been designed by that masterful Italian stylist, Giorgetto Giugiaro of Italdesign. It has a muscular stance, with a body made up of "emotional" surfaces—more normally associated with Jaguar, or Aston Martin—that tug at the heartstrings.

The Spyder has the distinctive, sleek presence of the established 3200GT coupe, yet its wheelbase is 220 mm (9 ins) shorter. The headlamps are amorphous when viewed head-on. The classic, jaw-like Maserati grille dominates the front, and the huge hood—reflecting pools of light—sweeps back to cover the powerplant. The doors waist slightly, and the wheel arches are gently flared to emphasize the gorgeous, fifteen-spoke wheels.

The fully insulated convertible should encourage owners to use their Spyders for touring in all weathers. When it needs to be folded, it can be automatically dropped at the press of a center-console button. It vanishes into a closed compartment (separate from the trunk) behind the seats.

Visible behind the seats are the fixed rollover hoops. True, they offer protection if the car overturns, but they're visually intrusive. Why couldn't Maserati have fitted an automatic system as on other, less expensive, convertibles? The interior is handcrafted in leather, and the seats are all-electric adjustable. Italian sports-car owners may be pleasantly surprised at the special care Maserati has taken to include functional storage compartments inside the cockpit. The trunk can hold two golf bags, or the optional set of fitted luggage.

This two-seat Maserati sports car has all the glamour of the company's famous 1960s models, such as the Ghibli and the Mistral, with none of their temperamental drawbacks. Only the $103,054 price tag should limit its appeal.

Matra m72

Engine	0.75 in-line 2
Power	37 kW (50 bhp)
Torque	61 Nm (45 lb ft) @ 3000 rpm
Gearbox	Automatic CVT
Front suspension	Double wishbone
Rear suspension	MacPherson strut, double wishbone
Brakes front/rear	Discs/discs
Front tires	155/80R15
Rear tires	185/80R15
Length	3240 mm (127.6 ins)
Width	1560 mm (61.4 ins)
Wheelbase	2230 mm (87.8 ins)
Track front/rear	1360/1370 mm (53.5/53.9 ins)
Curb weight	430 kg (948 lb)
0–100 km/h (62 mph)	14 sec.
Top speed	130 km/h (81 mph)

Matra previewed the production version of its m72 at the Geneva International Motor Show in March 2002. It is a recreational leisure car intended for the European market, one that mixes car and motorcycle for open-air fun in back-to-basics driving.

The exterior is minimalist, a combination of aluminum and composites. The functional front provides protection with the bumper, the windshield and the tire-hugging cycle fenders. The m72 looks—and is—light, at just 430 kg (948 lb) The visible aluminum beams are exoskeletal. The whole exterior suggests that functionality above anything else has achieved this lightweight goal. The rising door beam and the hoop that runs around the trunk lid simply emphasize the m72's nimbleness.

Aluminum and black materials are used extensively inside, giving a technical feel. Motorcycle design is referenced: for example, a red button starts the engine and plain, round instrument dials sit centrally. The general simplicity and openness of the cockpit, with its weather-resistant properties, should make the m72 a car to revel in.

This is the first time Matra will have produced a car under its own name in twenty years. In earlier decades its sports cars, such as the Djet, Bagheera and Murena of the 1960s, 1970s, and 1980s respectively, made interesting alternatives to Italian and British rivals. Its 1978 Matra Rancho was a pioneering leisure-oriented production model. Since 1983 Matra's principal automotive project has been manufacturing the Espace for Renault, but it will not build the next model. Now, with the m72, the French company aims to develop its own branded product while continuing to offer its engineering expertise to the automotive-industry giants, as in the case of its contract to build Renault's Avantime.

MATRA
m72

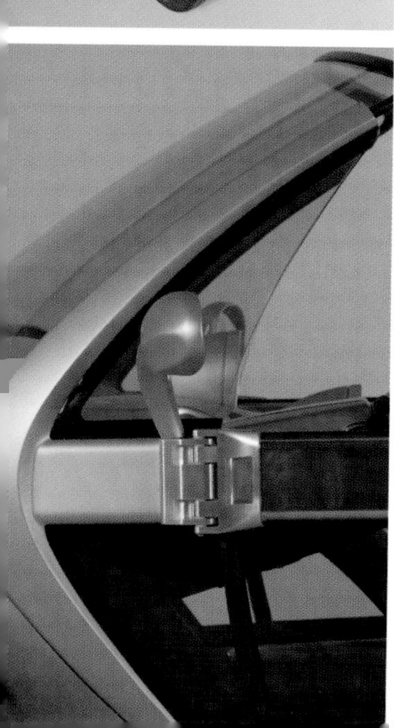

Maybach

The return of Maybach is one of the most vaunted auto-industry comebacks of recent times. It is, after all, a brand that has been dormant for sixty years. Until the Second World War, Maybachs were hand-built on the basis of individual customer specification. The ultimate model was probably the Zeppelin, with its V12 engine and awesome 5.5 m (18 ft) wheelbase which made it the largest German passenger car of its era. The Maybach name is being revived as a completely separate brand by DaimlerChrysler. Moreover, it expects Maybach to be, from the start, at the pinnacle of the global luxury-car market. Quite some undertaking.

The Geneva International Motor Show in 2002 was the venue for this über-prestigious rebirth. Harking back to the 1930s, the car's two-tone exterior paint offers hundreds of possible combinations, so each owner can choose a completely individual look.

Inside, the rear passengers enjoy acres of space and totally individual seats. These can be adjusted into a reclining position with automatically extending leg and foot supports—similar to those in first-class aircraft seating. High levels of interior technology are included: for instance, the entertainment and information center at the back has two flat-screen monitors linked to a TV receiver and a DVD player.

DaimlerChrysler aims to establish Maybach at the top of a small and highly distinguished market sector, one that experts believe holds further potential for growth over the next few years.

During the development phase, the design team canvassed potential customers around the world for their views. Maybach claims to have already a clear idea of what its target customers want. While they will be far more likely to be chauffeured in a Maybach than actually drive it themselves, technical perfection and stylish elegance share top priority with exceptional customer service.

Design	Stephen Mattin
Engine	5.5 V12 twin-turbo
Power	410 kW (550 bhp) @ 5250 rpm
Torque	900 Nm (664 lb ft) @ 2300 rpm
Installation	Front-engine/rear-wheel drive
Length	5720 or 6160 mm (225.2 or 242.5 ins)
Wheelbase	3830 mm (150.8 ins)

S MM 1244

Mazda6/Atenza

Design	Iwao Koizumi
Engine	2.3 in-line 4, for the 4- and 5-door models (1.8 and 2.0 in-line 4 also offered)
Power	122 kW (164 bhp) @ 6500 rpm
Torque	207 Nm (153 lb ft) @ 4000 rpm
Gearbox	4-speed automatic
Installation	Front-engine/front-wheel drive
Front suspension	Double wishbone
Rear suspension	Multi-link
Brakes front/rear	Discs/discs
Front tires	215/45ZR17
Rear tires	215/45ZR17
Length	4680 mm (184.3 ins)
Width	1780 mm (70.1 ins)
Height	1435 mm (56.5 ins)
Wheelbase	2675 mm (105.3 ins)
Track front/rear	1540/1540 mm (60.6/60.6 ins)
Curb weight	1415 kg (3120 lb)
0–100 km/h (62 mph)	8.9 sec.
Top speed	214 km/h (133 mph)
Fuel consumption	8.9 ltr/100 km (31.7 mpg)
CO_2 emissions	212 g/km

Mazda's far-reaching aim for its Mazda6/Atenza is that it should be the new global benchmark for mid-size cars in every aspect of styling, dynamic performance, packaging, quality, and taste. More than that, it must completely embody Mazda's new design mantra of "Emotion in Motion"—no mean feat when the product in question is a mainstream four-door sedan likely to be produced in substantial quantities.

The athletic design of 6/Atenza's color-coded exterior is a good start. At the front the air intake, headlamps and grille are low and wide to emphasize the car's width. The side profile is conventionally proportioned and solid-looking, with sporty details such as bumper and sill moldings, a rear-mounted spoiler, sports wheels and twin exhausts. It's certainly a progressive-looking car—especially next to Mazda's stylistically lackluster previous attempt with the 626—but it doesn't have the design vision of, for example, the new Renault Laguna.

Mazda claims to have raised the level of interior craftsmanship. Switches are said to have a pleasing tactility, and individual trim panels and parts are accurately fitted together. Ergonomic analysis conducted on target customers is said to have influenced the layout of controls and switches to achieve what Mazda believes is optimum functionality. The rear seats have an innovative folding mechanism inspired by *karakuri*, a traditional Japanese wind-up doll with a wide range of mechanical movements—for the perfect seating position.

The 6/Atenza isn't a bad-looking car. Its name is derived from the Italian word "*attenzione*," but it will probably be its specification—one that includes a 2.3 ltr engine, large alloy wheels, and double-wishbone suspension—that grabs the attention of customers, rather than any particular element of its inoffensive looks.

Atenza

Mazda MX Sport Runabout

Design	Moray Callum
Engine	1.5 in-line 4
Installation	Front-engine/front-wheel drive
Front tires	215/45ZR17
Rear tires	215/45ZR17
Length	3900 mm (153.5 ins)
Width	1720 mm (67.7 ins)
Height	1550 mm (61 ins)
Wheelbase	2500 mm (98.4 ins)

The MX Sport Runabout inherits Mazda's new brand and product DNA. It boasts athletic styling and has the current "family face," with its five-point air-inlet aperture and distinctive light clusters that echo the Mazda6/Atenza.

The exterior graphically emphasizes the cabin area while maintaining Mazda's surface language. Short overhangs and large-diameter wheels positioned well into the corners of the body help the MX to convey a clear sense of stability. To ram home the fun-to-drive message, the concept model's body is finished in vivid "yamabuki" orange.

The interior color is light beige, with chocolate lowlights on the upper instrument panel and door trim. Together with the metal-coated audio panel and door-trim insert, this enhances the chic and sporty design. Using architectural design principles, the interior emphasizes an efficient use of space while looking comfortable and inviting. This is achieved through Moray Callum's careful balance of form, materials, textures, and colors.

The Sport Runabout looks like a production-ready car. It certainly sits comfortably with Mazda's "Zoom Zoom" advertising tag line.

Atsushi '02
Atsushi '02

Mazda RX-8

Design	Yoichi Sato
Engine	2.4 rotary 2-rotor
Power	184 kW (247 bhp) @ 8500 rpm
Torque	220 Nm (162 lb ft) @ 7500 rpm
Gearbox	6-speed manual
Installation	Front-engine/rear-wheel drive
Front suspension	Double wishbone
Rear suspension	Multi-link beam
Brakes front/rear	Discs/discs, ABS
Front tires	225/45ZR18
Rear tires	225/45ZR18
Length	4425 mm (174.2 ins)
Width	1770 mm (69.7 ins)
Height	1340 mm (52.8 ins)
Wheelbase	2700 mm (106.3 ins)
Track front/rear	1500/1510 mm (59.1/59.4 ins)

It sometimes seems that there is a yawning chasm between concept and production cars. The new Mazda RX-8 bridges it. It is a sporty four-seater bristling with interesting and innovative technology and design elements—and it will actually be in Mazda showrooms any time now. It truly represents the once-troubled Japanese brand's new "Emotion in Motion" design philosophy.

In developing the RX-8, Mazda strove for conflicting goals: delivering a car with striking, sporty styling plus excellent handling and performance, while providing riding comfort for four adults. Mazda's designers have sculpted a kinetic body. From the front air intake to the truncated trunk lid, its surfaces are designed to be exciting and dynamic.

The rear-wheel-drive layout—like the RX-7, discontinued some years ago, the RX-8 uses a Wankel rotary engine for performance and smoothness—places visual emphasis on the rear tires and short front and rear overhangs, helping to underline the car's appearance of rock-solid stability.

This sports coupe is highly unusual in having four doors. They are conventional at the front, but the rear doors are radical: short, aluminum-made, and hinged at the rear. There is no central pillar, so, when both front and rear doors are open, access to the two rear seats is unusually easy.

The interior has strong, contoured surfaces that flow from the power bulge on the hood, through the dash and the center console, and right back to the rear parcel shelf. That console has a characteristic aluminum frame that echoes the form of the center "backbone" underneath it.

Up-to-the-minute sports-car styling, a radical rear-door concept, an interior package for four adults, and the world's only production automotive rotary powertrain: the RX-8 is an exciting car … and you can actually own one!

Mazda Secret Hideout

Engine	1.3 in-line 4
Power	65 kW (87 bhp) @ 6000 rpm
Torque	125 Nm (92 lb ft) @ 3500 rpm
Gearbox	4-speed automatic
Installation	Front-engine/rear-wheel drive
Front suspension	MacPherson strut
Rear suspension	Torsion beam
Brakes front/rear	Discs/discs
Front tires	185/65R15
Rear tires	185/65R15
Length	3925 mm (154.5 ins)
Width	1680 mm (66.1 ins)
Height	1530 mm (60.2 ins)
Wheelbase	2490 mm (98.03 ins)
Track front/rear	1480/1450 mm (58.3/57.1 ins)

Most car designs ultimately seek some sort of socio-economic statement about their intended customers. The bizarrely named Secret Hideout from Mazda, however, tries to tear up that rulebook. Its benign, friendly appearance, together with its warm and simple interior, hints at its aspirations to be something of a youthful automotive retreat from the world of economic pecking orders and executive-oriented aggressiveness.

The Secret Hideout doesn't try to be an upmarket car, despite its three small portholes which suggest, in a gimmicky way, limousine detailing. The styling is clean-cut, yet playful and distinctive. It features a solid, two-box package with upright windows, intended to maximize interior space. The blunt front end is stripped of any unnecessary features. It stands 1.53 m (5 ft) high but, to moderate this "tall-boy" look, the belt line is high, the passenger side windows are narrow, and the body's edges are smoothly rounded.

The exterior's color flows into the interior and is meant to be relaxing, with soft tones creating a gentle, cosy, living-room-like environment. The front and rear bench seats can be folded completely flat to create a bed. Both also feature plain lines—like pieces of modern furniture—to enhance the cabin's tidy appearance. The downside of this simplicity is that the seats appear to lack any lateral back support, and would no doubt become pretty uncomfortable after a short distance.

The car also features an engine starter button and push-button automatic transmission, both of which increase the aura of relaxed driver enjoyment.

There's nothing quite like the Secret Hideout on sale at the moment. If Mazda could confirm the demographic of laid-back youngsters as serious, then it could have an unlikely winner on its hands. For now, though, this remains strictly a concept car.

mazda
secret
Hideout

MCC Smart Crossblade

Engine	0.6 in-line 3 turbo
Power	52 kW (70 bhp) @ 5470 rpm
Torque	108 Nm (74 lb ft) @ 3500 rpm
Gearbox	6-speed sequential automatic
Installation	Rear-engine/rear-wheel drive
Front suspension	Wishbone
Rear suspension	Wishbone
Brakes front/rear	Discs/drums, ABS, EBD
Front tires	195/40R16
Rear tires	215/35R16
Length	2622 mm (103.2 ins)
Width	1618 mm (63.7 ins)
Height	1508 mm (59.4 ins)
Wheelbase	1812 mm (71.3 ins)
Track front/rear	1282/1393 mm (50.5/54.8 ins)
Curb weight	740 kg (1632 lb)
0–100 km/h (62 mph)	17 sec.
Top speed	135 km/h (84 mph)
Fuel consumption	5.7 ltr/100 km (16.1 mpg)
CO_2 emissions	137 g/km

One short year after the concept for the car was shown at the Geneva International Motor Show in 2001, MCC Smart now offers a production version of its radical Crossblade.

Lacking a roof, doors or conventional windshield, it is designed to demonstrate just how far Smart is prepared to go with a radically pared-down model. Not since the Mini sired the stark four-seater Moke utility in 1964 has so little bodywork been offered on a production car. Only two thousand Crossblades will be built in MCC's factory in northern France, each individually numbered to emphasize its exclusivity.

Smart says that this unconventional design gives the sensation of riding a four-wheeled motorcycle. A narrow, tinted wind deflector sits ahead of the driver and passenger to direct the rushing air over their heads, while a sturdy rollover-bar behind the seats provides protection in the unfortunate event of a flip-over. Smart claims that the safety standards of the Crossblade are comparable to those of its Cabrio and City Coupe models, owing to additional floor reinforcements.

The lack of a roof means that the interior is designed to resist the elements. The instrument panel and seats are covered with water-repellent red plastic, and a plastic floor liner has four drainage holes to carry rainwater away. The Crossblade also looks unique, with its black front and rear ends, and prominent matt-black front and rear spoilers.

This is a perfect car for carefree drivers, ideally in a Mediterranean climate. Sunglasses—plus, of course, a bubble jacket—are a must.

smart
BB X 252

MCC Smart Tridion4

The original ultra-compact Smart City Coupé, 2.5 m (8.2 ft) in length, is no easy act to follow, particularly when the concept is stretched into a four-seat, five-door, family-car rival to European market heavyweights such as the Ford Fiesta and Volkswagen Polo. That, however, is what the Tridion4 is trying to achieve.

Smart's designers have elongated the strict one-box profile of the City Coupé, while still maintaining the distinctive face with high-mounted headlamps, air inlets, and service grilles typical of the brand. The prominent fender moldings extend to cover the wheels, and are colored to match the doors and contrast with the body. At the rear, the large glass tailgate wraps around and cradles high-mounted rear lamps.

Why bother? Smart, part of industry giant DaimlerChrysler, needs to expand its brand "reach" while using shared platforms, most notably that of a small car to come from another DaimlerChrysler affiliate, Mitsubishi. Squaring up to more conventional big hitters such as the Polo will inevitably neutralize some of Smart's more characterful attributes.

Still, the use of different colors and materials for the body exterior emphasizes the form and extent of the separate panels, as in the City Coupé. The safety passenger cell is steel, and is visible on the front- and rear-window pillars, the roof side structure and the door sills. The front and rear bumpers and doors are clad with deformable, scratch-resistant plastic panels. The large glass panel spanning the entire roof (another Smart hallmark) has a blue tint, designed to create a light and comfortable interior. The leather seats have an innovative feature: when the front ones are folded, they can be converted into a sofa.

Smart will have a four-door car on sale within months. Whether or not it looks much like Tridion4, it will have to work hard at grabbing a market share.

Length	3650 mm (143.7 ins)
Width	1790 mm (70.5 ins)
Height	1450 mm (57.1 ins)

smart
TRIDION 4

Mercedes-Benz CLK

Design	Jurgen Bollmann
Engine	5.5 V8 (1.8 in-line 4, 2.6 and 3.2 V6, 5.0 V8, and 2.7 in-line 5 diesel also offered)
Power	270 kW (362 bhp) @ 5750 rpm
Torque	510 Nm (376 lb ft) @ 4000 rpm
Gearbox	5-speed automatic
Installation	Front-engine/rear-wheel drive
Front suspension	Multi-link with MacPherson strut
Rear suspension	Multi-link
Brakes front/rear	Discs/discs
Front tires	225/40R18
Rear tires	255/35R18
Length	4638 mm (189.7 ins)
Width	1740 mm (68.5 ins)
Height	1414 mm (55.7 ins)
Wheelbase	2715 mm (106.9 ins)
Track front/rear	1495/1474 mm (58.9/58 ins)
Curb weight	1715 kg (3804 lb)
0–100 km/h (62 mph)	5.2 sec.
Top speed	250 km/h (155 mph)
Fuel consumption	12.8 ltr/100 km (18.3 mpg)
CO_2 emissions	276 g/km

The Mercedes-Benz CLK already has an illustrious history. This new model features an entirely new exterior and interior design, together with plenty of new technology.

At the front, inspiration is drawn from the CL-Class—the bigger brother of the CLK coupe—and the SL roadster, both of which acted as templates for the classic radiator grille with its distinctive vanes and large, central, Mercedes star emblem. The twin-headlamp "face" comprises two ellipses merged to form a single headlamp unit. Clear glass conceals four separate lighting systems.

These headlamps set a formal tone at the front: the hood and front fenders follow the rounded contours, continuing rearward along the body. Slim pillars extend up on to the sweeping roof line, which continues into the C-pillar, sloping gently into the coupe rear. Indeed, the new CLK is best viewed from the side. There is no upper door pillar, so when all the side windows are dropped you can get a clear view of the interior.

The suave exterior form is echoed inside. The dash uses a combination of straight lines and smoothly rounded contours to convey lightness, dynamism, and elegance. Sporty, round air outlets sit in the dash, chrome surrounds the dials, and generous servings of wood or aluminum trim adorn the center console.

Technology available on the CLK now includes "Distronic," an autonomous, intelligent cruise control, "Keyless-Go" access and drive authorization system, bi-Xenon headlamps, and "Linguatronic" voice-operated control of phone and audio.

Mercedes-Benz E-Class

Design	Hans-Dieter Futschik
Engine	5.0 V8 (2.6 and 3.2 V6, 2.2 in-line 4, and 2.7 in-line 5 turbo-diesel also offered)
Power	225 kW (306 bhp) @ 5600 rpm
Torque	460 Nm (339 lb ft) @ 2700–4250 rpm
Gearbox	5-speed automatic
Installation	Front-engine/rear-wheel drive
Front suspension	4-link suspension
Rear suspension	Multi-link independent suspension
Brakes front/rear	ABS, ESP, SBC
Front tires	245/45R17
Rear tires	245/45R17
Length	4818 mm (190 ins)
Width	1822 mm (71.7 ins)
Height	1430 mm (56.3 ins)
Wheelbase	2854 mm (112.4 ins)
Track front/rear	1567/1560 mm (61.7/61.4 ins)
Curb weight	1725 kg (3803 lb)
0–100 km/h (62 mph)	6 sec.
Top speed	250 km/h (155 mph)
Fuel consumption	11.5 ltr/100 km (20.4 mpg)
CO_2 emissions	276 g/km

Mercedes-Benz reveals that it took forty-eight months to develop the new E-Class. In the process it invested over € 2,000,000,000 in this, its staple and core executive-car range. For that commitment you should expect a lot of changes, yet traditional E-Class buyers will want none of the car's established qualities tampered with. Quite a challenge.

The new E-Class design inherites the elegant lines of the many earlier series, but adds a dynamic flavor. The twin headlamps with which the E-Class made its mark in 1995, initiating a design trend, are redesigned and slightly angled back, characterizing the progressive appearance of the car's nose treatment.

Despite the obvious design changes, the E-class focus has mainly been on technical innovation. Vehicle safety and handling dynamics are where the main investment has been made, such as the electro-hydraulic braking system, Sensotronic Brake Control (SBC), and the Electronic Stability Program (ESP), pioneered by Mercedes-Benz, which is standard E-Class equipment. The enhanced Airmatic Dual-Control (ADC) air-suspension system improves ride comfort, and is now also included.

Inside is a new multi-contour seat that adapts to the driving environment. Its air chambers automatically inflate and deflate to offer the driver and front passenger the best possible lateral support.

Available as an optional extra is the four-zone Thermotronic climate control, whose microprocessor individually controls temperatures for the driver and front passenger, as well as the passengers on the outer rear seats. For the first time, the navigation system in the new E-Class can also be controlled using the "Linguatronic" voice-control system.

The E-Class is now better than ever, and Mercedes-Benz can confidently expect good sales to its loyal followers.

S NP 6072
S NP 6072

Mercedes-Benz F400 Carving

Design	Steffan Kohl, Alexander Buckan, and Patrick Reimer
Engine	3.2 V6
Power	160 kW (218 bhp)
Torque	310 Nm (229 lb ft)
Gearbox	Sequential automatic
Installation	Front-engine/rear-wheel drive
Front suspension	Active hydro-pneumatic
Rear suspension	Active hydro-pneumatic
Brakes front/rear	Discs/discs, brake-by-wire, SBC
Front tires	255/35R17
Rear tires	255/45R19
Length	3979 mm (156.7 ins)
Width	1890 mm (74.4 ins)
Height	1150 mm (45.3 ins)
Wheelbase	2450 mm (96.5 ins)
Track front/rear	1586/1586 mm (62.4/62.4 ins)
0–100 km/h (62 mph)	6.9 sec.
Top speed	241 km/h (150 mph)

Despite its deliberately radical appearance, Mercedes-Benz claims that its F400 Carving is more than a mere concept car. It is, say its German creators, a mobile research laboratory brimming with technical innovation. You might assume that you will never be able to buy one, but recent Mercedes showroom models, such as the A-Class and C-Class coupe, have also proved difficult to pigeonhole.

The F400 Carving's outrageous lines hide leading-edge technology. It is a stripped-down speedster with an elongated, low-slung hood, a short tail, and an interior purpose-built for two. Its high-performance looks are reinforced by the wide, low-slung air intake in the front section, racing-car-like omission of a windshield, widely set exhaust pipes, and distinctive rollover bars for driver and passenger.

Wing-like sections powerfully spanning the wheels, and smaller wing sections fore and aft of the wheels, grab attention when the car is viewed from the side. However, the most striking elements are its gull-wing doors, a historic feature pioneered by Mercedes on the 300SL exactly fifty years ago.

Mercedes-Benz is a byword for automotive safety. The Carving features active computer-controlled camber adjustment, together with specially developed tires that generate an extra 30% lateral grip over conventional systems. This gives higher possible cornering speeds and enhanced active safety.

In the interior, the transmission tunnel has the shape, color, and texture of a cast-aluminum transmission bell housing, echoing Mercedes-Benz racing-car cockpits of the 1920s and 1930s. The dash is fitted with simple metallic controls, designed to reinforce the retro image. Yet behind each function lies modern technology.

The F400 Carving bears comparison with the extreme Lotus 340R in its styling and outright performance. The Mercedes is technologically advanced in comparison with the Lotus's raw thrills, but the 340R proves that such a pared-down car can be a roadgoing reality.

Mercedes-Benz SL

Design	Stephen Mattin
Engine	5.0 V8
Power	225 kW (306 bhp) @ 5600 rpm
Torque	460 Nm (339 lb ft) @ 2700–4200 rpm
Gearbox	5-speed automatic
Installation	Front-engine/rear-wheel drive
Front suspension	4-link, Active Body Control
Rear suspension	Multi-link independent, Active Body Control
Brakes front/rear	Discs/discs, ABS, Brake Assist, ESP, SBC
Front tires	255/45R17
Rear tires	255/45R17
Length	4535 mm (178.5 ins)
Width	1815 mm (71.5 ins)
Height	1298 mm (51.1 ins)
Wheelbase	2560 mm (100.8 ins)
Track front/rear	1559/1547 mm (61.4/60.9 ins)
Curb weight	1845 kg (4068 lb)
0–100 km/h (62 mph)	6.3 sec.
Top speed	250 km/h (155 mph)
Fuel consumption	12.7 ltr/100 km (18.5 mpg)
CO_2 emissions	304 g/km

The new SL is the fifth generation of this incredibly successful series from Mercedes-Benz, which began with the classic gull-wing-door 300SL in 1954. That first car was a technical tour de force—the first four-stroke production car in the world with fuel injection—but it was also a race-bred sports car. Starting with the 230SL in 1963, the cars became progressively softer and more civilized. Now the new SL, in design terms at least, is trying to recapture the edgy nature of the original.

Compared to the previous model, the new SL has a crisper side profile and a more steeply raked-back windshield. Clear headlamps are used, and indicators are mounted in the exterior mirrors, like other recent Mercedes models. Chrome-laced air intakes on the front fenders summon up vivid recollections of the 300SL. While SLs have long come with a removable hard-top, the new SL has adopted a folding Vario-roof system similar to that of the smaller SLK roadster. Combining the open-air pleasure of a roadster with the comfort of a Mercedes coupe, at the push of a button the roof opens or closes in sixteen seconds. During that time, a new tilting mechanism ensures that all the roof components retract into the top section of the trunk. Luggage space is infringed by the process, so traditional SL buyers must travel a bit lighter. The interior uses leather, wood, and aluminum for that classic SL feeling of exclusivity and quality.

The SL is rich in new technology. Sensotronic Brake Control (SBC) communicates the driver's commands electronically via cable. If the driver switches his or her foot quickly from accelerator to brake pedal, SBC recognizes the early signs of an emergency situation. With the help of a high-pressure reservoir, the system automatically raises the pressure in the brake lines, instantly pressing the pads on to the brake discs, which then are ready to spring into action as soon as the brake pedal is pressed. This system could, says Mercedes-Benz, make all the difference in an emergency; expect to see it on mainstream Mercedes models soon.

The SL is the flagship Mercedes convertible. The practicality and quality of its new roof system, plus its desirable aesthetics, will secure its position as a future classic, even if buyers accustomed to previous SLs could find it just a touch too radical at first.

S SL 5535

Mercedes-Benz Vaneo

Design	Gerhard Honer
Engine	1.9 in-line 4 (1.6 in-line 4 also offered)
Power	92 kW (125 bhp) @ 5500 rpm
Torque	180 Nm (133 lb ft) @ 4000 rpm
Gearbox	5-speed manual
Installation	Front-engine/front-wheel drive
Front suspension	MacPherson strut, coil springs
Rear suspension	Trailing-arm axle, coil springs
Brakes front/rear	Discs/discs, ABS, EBD
Front tires	195/55R15
Rear tires	195/55R15
Length	4192 mm (165 ins)
Width	1742 mm (68.6 ins)
Height	1830 mm (72 ins)
Wheelbase	2900 mm (114.2 ins)
Track front/rear	1524/1477 mm (60/58.1 ins)
Curb weight	1375 kg (3032 lb)
0–100 km/h (62 mph)	10.8 sec.
Top speed	180 km/h (112 mph)
Fuel consumption	8.1 ltr/100 km (29 mpg)
CO_2 emissions	197 g/km

The Vaneo, new from Mercedes-Benz, is a true hybrid of family sedan, recreational vehicle, and utility wagon. It is aimed at those customers with a multitude of uses for their car. It shares much of its technology with the A-Class—the company's successful and unusual entrant into the Golf/Focus/Corolla family-car class—but with a new and very van-like upper architecture.

This is what the car industry calls a "monospace" or "one-box" design—just one unit for passengers, engine, and luggage, with no protruding hood or trunk. Its angular front is very similar to the A-Class's, and the large side windows with wide corner radiuses, a common feature of vans, give a low belt line, and an honest, utilitarian look. The Vaneo's compact hood makes for an eye-catching contrast with the large windshield. Teardrop-shaped, monochrome headlamps give the car a friendly "face." The radiator grille is placed between them, along with black fins and the integral Mercedes star emblem. A subtle ridge in the headlamp lenses is picked up by the top edge of the grille, and flows back along the sides of the vehicle as a bold contour, running parallel to the lower edge of the windows. The twin, sliding side-rear doors are unimpeded by the rear wheels, giving good access into the load area.

Three Vaneo customization packages are available: trend, family, and ambient. In addition, a range of recreational options gives plenty of scope for individuality.

The Vaneo is a highly practical car with the flexibility to cope with dozens of uses. Mercedes-Benz can be confident of its wide sales potential.

S GA 1241
S GA 2260

Mercedes-Benz Vision GST

Design	Stephen Mattin
Engine	5.5 V8
Power	265 kW (360 bhp)
Torque	530 Nm (391 lb ft) @ 3000 rpm
Gearbox	Automatic
Installation	Front-engine/4-wheel drive
Front suspension	AIRMATIC
Rear suspension	AIRMATIC
Brakes front/rear	Discs/discs, ABS, ESP
Front tires	275/45R22
Rear tires	295/40R22
Length	5150 mm (202.8 ins)
Width	1970 mm (77.6 ins)
Height	1690 mm (66.5 ins)
Wheelbase	3220 mm (126.8 ins)
Track front/rear	1670/1670 mm (65.7/65.7 ins)
0–100 km/h (62 mph)	7 sec.

It is a fair bet that history will judge the Mercedes-Benz Vision GST to be much more than the design concept that so perfectly captured the spirit of the 2002 Detroit show. Without doubt, the GST marks the debut of a whole new breed of "crossover" vehicles, so called because they combine the functions of today's luxury sedans, station wagons, people-carriers, and four-wheel-drive sport utilities.

At first glance, the GST comes across as a sleek sports coupe, but with its rear roof line extended to provide greater interior space. Step closer, however, and the sheer size of the 5 m (16.4 ft) long study becomes immediately apparent. In reality this is a large and versatile wagon, seating six people in three rows, that disguises its bulk through clever design.

The most dramatic of the design elements is the big, sweeping arch of the window line. It starts at the A-pillars, rises in a graceful curve to the roof, and sweeps back down to meet the rising belt line at the tailgate. This helps give a sporty profile from the side—a deceptive impression, as the main roof line goes its own way, staying higher in order to provide interior headroom. The roof itself is mainly glass, electro-chromatic so that passengers can alter its tint according to the intensity of the sun.

Rear-hinged rear doors mean no B-pillars to obstruct access to the roomy and innovation-packed cabin. The instrument panel is dominated by twin dials in front of the driver, and a futuristic navigation display that appears to hover in thin air in the center. Novel mood-lighting techniques add to the already luxurious interior ambience. The belt systems are integrated into the seats to improve convenience and save clutter.

Mechanically, the GST is based on the platform of the upcoming replacement for the M-Class sports utility. As such, it incorporates every advanced chassis and communication system known to the automobile world. If ever there were a concept clearly earmarked for production—and success—this is it.

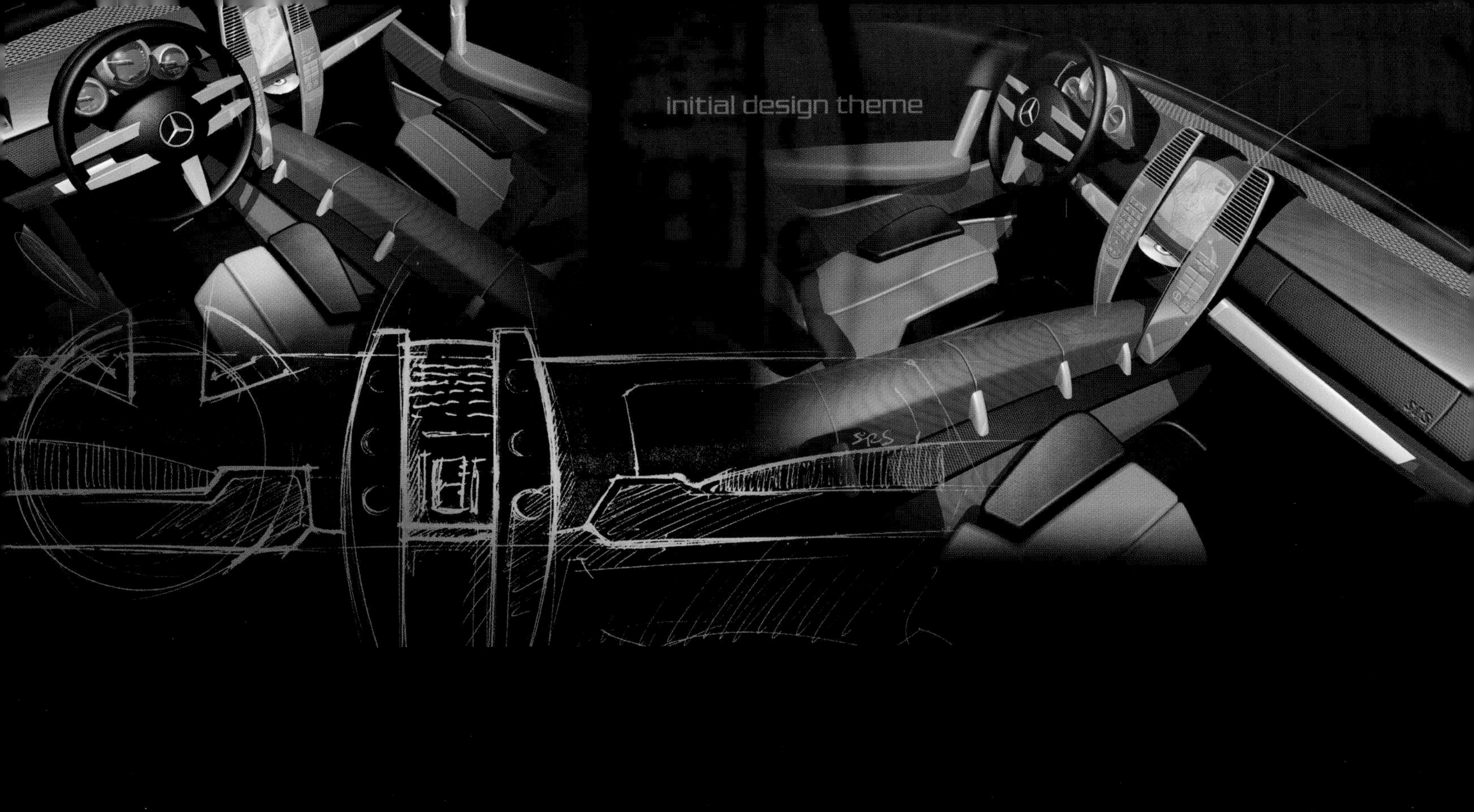

initial design theme

MG X80

Design	Peter Stevens
Engine	4.6 V8
Power	282 kW (385 bhp)
Gearbox	Manual or automatic
Installation	Front-engine/rear-wheel drive
Front suspension	Double wishbone
Rear suspension	Double wishbone
Brakes front/rear	Discs/discs

The MG X80 is planned for production in both coupe and convertible forms, and in double-quick time. This is because the basis of the car came ready-designed, after the newly independent MG Rover bought an obscure Italian manufacturer called Qvale in June 2001.

Qvale's main asset was the Mangusta, a two-seat, rear-wheel-drive sports car with a V8 Ford Mustang engine. This car began life as a De Tomaso, and had been inexpensively and cleverly developed to the point where it had just gone on sale in the US.

Perhaps surprisingly for an Italian car—especially one styled by Marcello Gandini, creator of the Lamborghini Countach's looks—the Mangusta's styling was overshadowed by its excellent performance. But MG Rover knew that the car's appearance could be overhauled by Peter Stevens, its head of design.

Its basis is a "superformed" aluminum body (similar in construction to the Aston Martin Vanquish), mounted on a steel box-section chassis. On this architectural structure, Stevens's distinctive MG grille sits ahead of a long hood, which also features two chrome motifs near the windshield. The side profile is undecorated, apart from a small step that runs at waist level and a vent that lies at the rear of the front fender. To emphasize its performance, the X80 is fitted with sill moldings, 18 ins wheels, a round aluminum fuel filler, and four exhausts. The interior—as yet unseen—will be a leather-lined cocoon.

The X80 retains the Mangusta's dynamic specification: independent wishbone suspension, a front-engine, rear-wheel-drive layout, and a variety of power outputs produced from 4.6 ltr V8 engines. Other performance features include electronic traction control, a limited-slip differential, and anti-lock braking.

The two-plus-two-seater X80 coupe will be launched first, giving the MG brand a product in the high-performance, luxury-sports-car sector of the market for the first time since the demise of the MGB GT V8 in 1976. The roadster version will follow later. Producing an expensive and luxurious car quickly is a design and engineering feat. However, the X80 has a tough assignment—it must elevate the MG brand into profitable new territory.

MG X80

MG X80

Mitsubishi CZ2

Design	Olivier Boulay
Engine	1.3 in-line 4
Gearbox	Automatic CVT
Length	3830 mm (150.8 ins)
Width	1695 mm (66.7 ins)
Height	1475 mm (58.1 ins)
Wheelbase	2500 mm (98.4 ins)

The CZ2 from Mitsubishi is a proposal for a new small hatchback—a market sector in which Mitsubishi has consistently failed to shine. However, its chic looks and single-motion silhouette compare favorably with today's Alfa Romeos and Peugeots.

The CZ2 has a soft, almost comforting shape, interrupted by dynamic features that give the car a vibrant persona. The rounded front end has dramatically pointed headlamps that cut into its hood, while the windshield extends up into the roof to provide a bright and spacious interior, even if that makes for a dazzling view of the road ahead in bright sunshine.

The doors are gently curved, with a small, sharp feature line that runs lengthways through the horizontally mounted door handles. The fenders protrude subtly to cover the wheels, giving the CZ2 a sporty, agile appearance. The interior has a wave-shaped instrument panel, and some convenience-enhancing features, such as a bench-type front seat, detachable door-trim pouches, and a large, four-part glass roof with independent sunlight control, creating a desirable ambience.

It's undeniable that the exterior design of the CZ2 makes clear reference to other production models, including Peugeot's 206, Alfa Romeo's 147 and the new Seat Salsa—all of which use rounded, harmonious shapes as part of their market-pleasing packages. The side profile, headlamps, and detailed body features all possess an appealingly Mediterranean quality. This is a world away from the underwhelming image of Mitsubishi's current small hatchback, the Colt. The CZ2 could be a desirable car, particularly for Europeans, and a big leap forward for its maker.

Mitsubishi Pajero Evolution 2+2

Design	Olivier Boulay
Engine	4.7 V8
Gearbox	6-speed sequential, permanent AWD
Installation	Front-engine/4-wheel drive
Front tires	285/45R22
Rear tires	285/45R22
Length	4420 mm (174 ins)
Width	1980 mm (78 ins)
Height	1725 mm (67.9 ins)
Wheelbase	2780 mm (109.5 ins)
Track front/rear	1695/1695 mm (67.7/67.7 ins)

Jutta Kleinschmidt, winner of the 2001 Paris–Dakar Rally, has been consulted throughout the design of the Mitsubishi Pajero Evolution 2+2 concept, bringing her motorsport expertise into play. More women designing cars has to be a good thing, but Kleinschmidt also brings some welcome credibility to what might otherwise be perceived as just another brawny auto-show diversion.

The Evolution is a car conceived by Mitsubishi's new chief designer, Olivier Boulay, specifically for competition rallying. His new corporate design brief calls for greater visual emphasis, while maintaining Mitsubishi's 4×4 heritage and technical excellence.

The Evolution has an intimidating, chiselled design well suited to its prime objective—its functionality. High ground clearance, huge wheels, and extended wheel arches should mean rapid progress over rough terrain. Other no-nonsense features include the large tailgate-mounted wing, the sill steps, and the finely meshed grilles covering much of the front, maintaining the flow of cooling air while keeping out stones. Between the front grilles is Mitsubishi's trademark, its enlarged three-diamond emblem.

Power for the Evolution comes from a 4.7 ltr V8 engine. To ensure that Mitsubishi remains competitive in auto sport, the Evolution is fitted with a super-select four-wheel-drive system, Active Yaw Control (AYC), and an Active Center Differential (ACD).

Form is definitely subordinate to function in the Pajero/Shogun Evolution. But maybe Mitsubishi would be wise to ensure that functionality is not pursued at the expense of aesthetics in future mainstream models.

V8

V8

Mitsubishi Space Liner

The Space Liner is another radical Mitsubishi concept: a futuristic, one-box car designed to carry four people in unashamed luxury. Although Mitsubishi admits it is just a design study for now, its creators have schemed a level of interior space normally found only in limousines. It is an imposing machine, long, low, and wide. Its ground-hugging profile is echoed in the thin band of body color running along the waist of its doors, and extending forward and backward over the wings.

The Space Liner proposes a new way of looking at the way passenger space is used in cars. A unique seating layout heads the list of its innovative interior features. It encourages its pampered occupants to stretch out and relax, offering generous head, leg, and shoulder room for up to four adults.

Also featured are touches of Japanese domestic interior design, from indirect lighting to translucent sun shades, to create an unusually relaxing environment. Features designed to reduce stress include front and rear side doors that open wide from the center pillar, facilitating easy entry and exit, revolving front power seats, and a soft-shaped rear seat.

The Space Liner allegedly uses advanced technology. A "lane-trace assist" function would help alleviate driver stress by keeping the car within its lane. A drive-by-wire system connects accelerator and brake controls to the steering wheel. Another part of the Space Liner's intended technical package is fuel-cell power, connected via a four-wheel-drive system comprising independent front and rear motors.

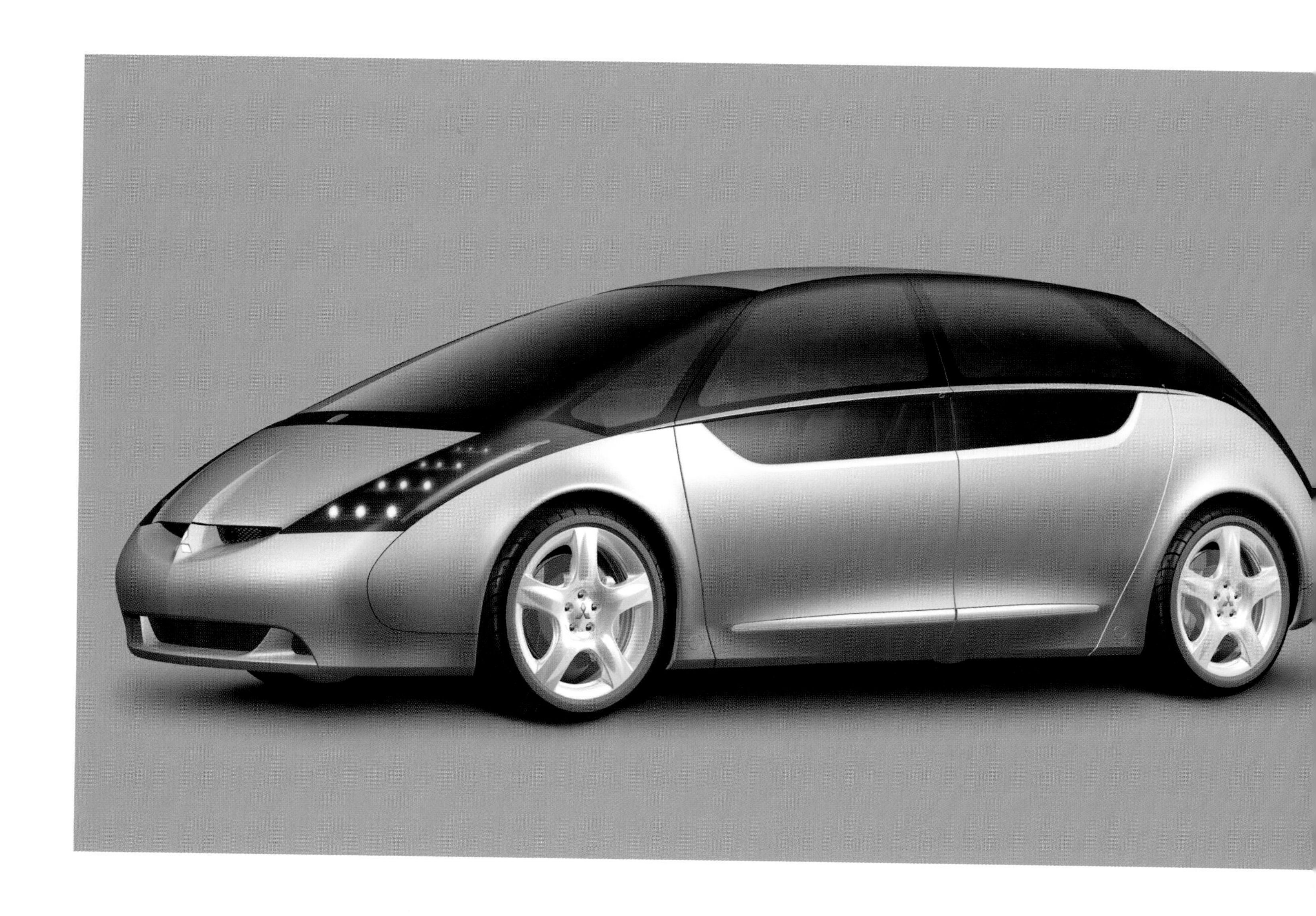

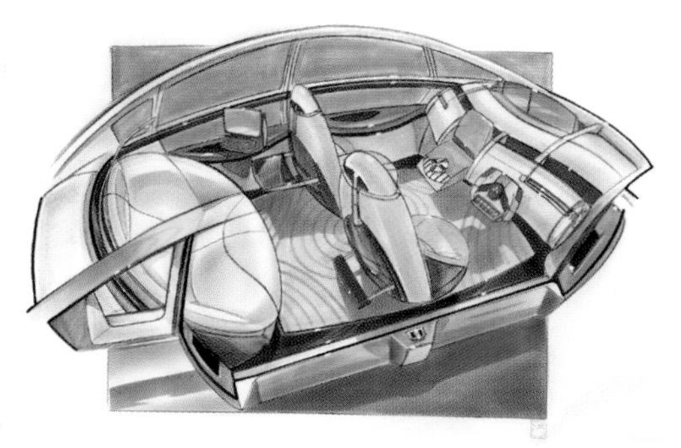

Mitsubishi SUP

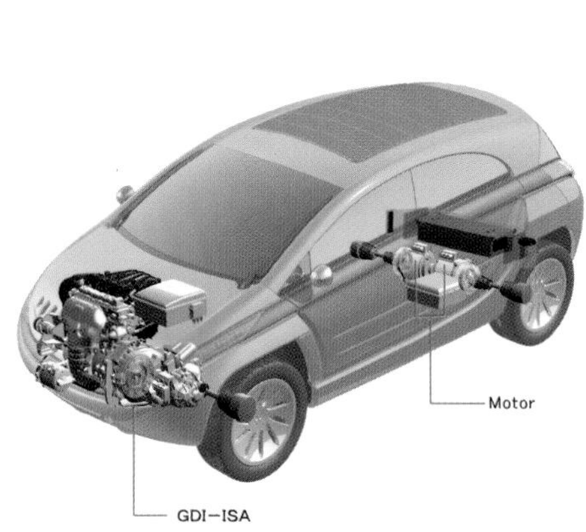

According to Mitsubishi, today's society is full of consumers who want "actively [to] take advantage of hi-tech equipment in their everyday lives, and the offerings of the great outdoors in their free time." This is the thinking behind its Sports Utility Pack (SUP) concept, unveiled at the Tokyo Motor Show in October 2001, with the Cabrio being shown at Detroit in January 2002. It should be equally at home in the city or the country, bridging the gap between Mitsubishi's conventional sedans, such as the Lancer, and its thoroughbred off-roaders, such as the Pajero and Pinin.

Essentially a mono-form broken up by strong design elements, the car is playful and fun. The SUP's front has slotted air vents in the hood, a sump guard containing the driving lamps, and a strong, tube-like band, containing futuristic-looking headlamps, draped over the car—and dominating it from every angle. Other details that make the SUP look youthful and active are different-colored door skins, eye-catching door mirrors, a large, louvered, power sun-roof, and detachable translucent door pockets.

The SUP's all-terrain capability comes from a drivetrain with an automated manual transmission that can apply the clutch automatically and change gear. Each rear wheel is driven by its own electric motor for four-wheel drive.

This is another well-executed concept from Mitsubishi, allying complex forms and color combinations with new technology. Perhaps a little too radical for production in its current form, the SUP would be much more feasible once its gimmicky design features have been eliminated.

SUP

Nissan Crossbow

Design	Christopher Reitz and Nissan Design Europe
Front suspension	Multi-link independent
Rear suspension	Multi-link independent
Front tires	20 ins PAX run-flat system
Rear tires	20 ins PAX run-flat system
Length	4860 mm (191.3 ins)
Width	1965 mm (77.4 ins)
Height	1830 mm (72 ins)
Wheelbase	3000 mm (118.1 ins)

The Nissan Crossbow is an exercise in design that combines a new look with a luxury interior in a serious, V8-powered off-roader. The Crossbow's shape is strongly emphasized by rectangular forms, and the thick lines that surround its doors, giving a feeling of depth and solidity to the steel compartment.

Nissan, of course, is no stranger to uncompromisingly "square" design; generations of its rugged, 4×4 Patrol have been resolutely bluff. The Crossbow, though, appears to have taken its inspiration more from America's military Hummer than any Japanese design tradition. Unique touches include the unconventional rear doors, rear-hinged to give improved access. When both side doors are opened, the sill drops down automatically to provide a step, as in an executive jet—neat, if a little unnecessary.

Other features designed for off-roading include large, straight-edged wheel arches, short front and rear overhangs, and chunky tires on huge 20 ins wheels. The front bumper has a removable panel in the center that conceals a built-in winch. Twin exhaust pipes fit flush with the rear bumper to keep them from getting snagged as the Crossbow climbs a hill. The wipers park vertically, and run on concealed tracks, so they can clean the entire windshield.

The interior forms a contrast to the rugged exterior. Natural materials such as olive wood are used to create a luxurious and calm atmosphere. The tan-colored dash has silver edges, and a T-shaped screen mounted in the center console that integrates the navigation, audio, and climate controls. Information on road surface, traction levels, gradient, and altitude is also given.

The Crossbow differs significantly from Nissan's Terrano and Patrol. Could its sharp edges and rectangular form be the new face of Nissan 4×4 design?

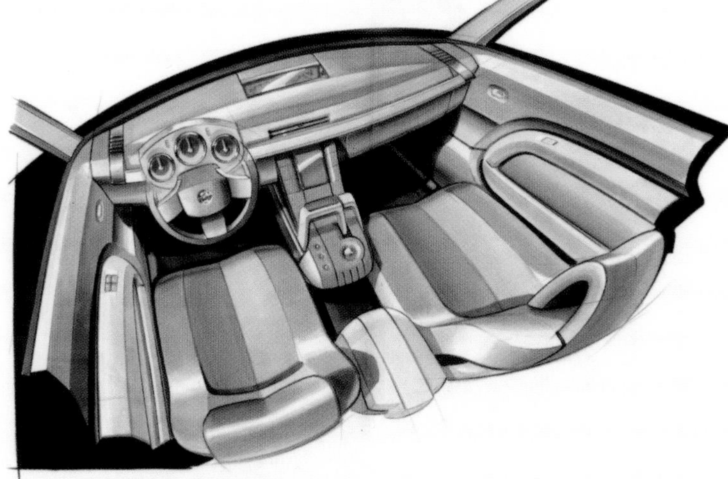

Nissan GT-R

The new GT-R from Nissan will probably soon replace the dated but legendary Skyline GT-R—a sports coupe with a revered reputation, garnered over many years, as a formidable track racer in Japan. The new car will remain just as extreme, as technically advanced, and as dedicated to driving pleasure as ever, but will be wrapped in a more stylish body.

The massive front air intake, for many years one of the GT-R's most striking visual characteristics, hints at this model's performance. A discreet power bulge at the trailing edge of the hood, near the windshield, adds to the effect, while vertically stacked headlamps that seemingly spear rearward—cutting a graphic V-shape into the tops of the front fenders—reinforce the messages of speed and potency.

Bulging wheel arches, rounded at the front and boldly wedge-shaped behind, stretch tautly over massive 19 ins wheels, creating svelte haunches. Unpainted carbon-fiber sill panels disguise large air ducts for cooling the rear brakes. Carbon fiber is also used for the substantial aerodynamic diffuser underneath the tail, for lightness.

A fore-and-aft stripe bisecting the GT-R's roof adds an image of strength and solidity to the whole car. The design of its two-plus-two interior is decidedly functional, dominated visually and physically by a central metal frame that splits the cabin. This is the core around which the whole car is built. Digital screens monitor the GT-R's vital functions, providing minimum data and maximum information so the driver's concentration remains unbroken.

Design	Makato Yamane

Nissan Ideo

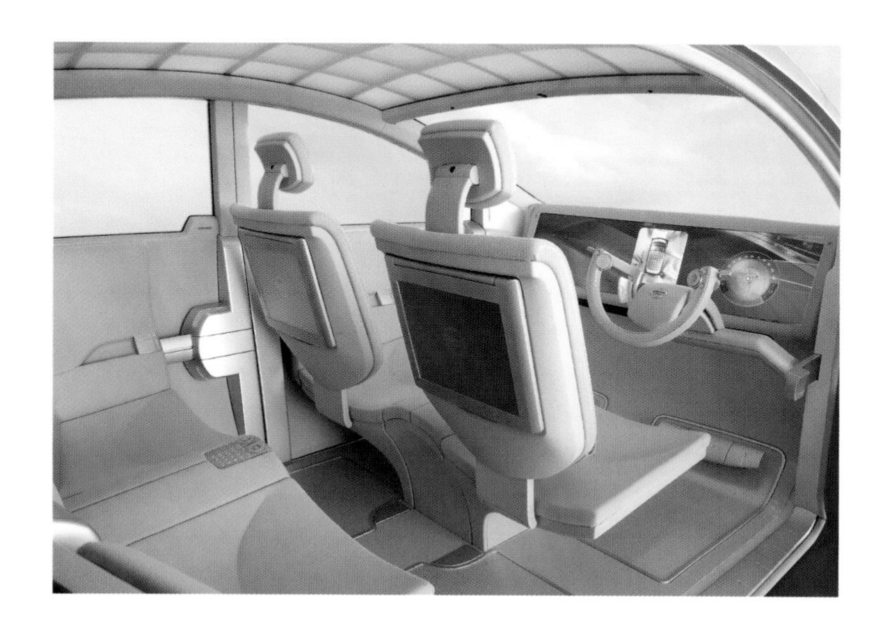

Design	Masato Inoue
Front suspension	Double wishbone
Rear suspension	Double wishbone
Front tires	175/55R17
Rear tires	175/55R17
Length	3600 mm (141.7 ins)
Width	1700 mm (66.9 ins)
Height	1620 mm (63.8 ins)
Wheelbase	2730 mm (107.5 ins)
Track front/rear	1325/1325 mm (52.2/52.2 ins)

Here is a forward-looking concept for a city car from Nissan—the Ideo—that incorporates some fascinatingly Japanese architectural design elements into its outwardly bluff appearance. The glass roof, with its aluminum lattice "ceiling," appears very rigid, while allowing light positively to drench the cabin. The lattice detail is echoed on the semi-translucent sides of the front headlamps which, believe it or not, Nissan claims are inspired by Japanese paper lanterns.

The cab-forward design uses large block surfaces, straight edges, and sharp radiuses for a contemporary feel. The interior is simple and modern, using bright colors mixed with aluminum trim that makes it look technically appealing and interestingly fresh.

The lack of side bolstering on the seats suggests that the Ideo is meant for gentle city driving; more spirited maneuvers would have the driver sliding around in his seat. Conventional instruments are replaced by a large screen in front of both driver and front passenger, displaying vehicle information, and relaying feedback from cameras ingeniously installed around the car. Large screens mounted on the back of the front seats provide entertainment for rear-seat occupants, who also have plenty of foot space thanks to the front seats' frameless design.

The Ideo is slightly larger than the more traditional Nissan Moco concept (see p. 184), and is much more imaginative from a design perspective, with its new concepts, and far more contemporary style. It would be a brave and inspired Nissan that incorporated the Ideo's architectural features into future products.

NISSAN
ideo

ideo

ideo

Nissan Kino

Design	Stephane Schwarz
Gearbox	4-speed automatic
Front suspension	MacPherson strut
Rear suspension	Multi-link beam
Front tires	18 ins PAX run-flat system
Rear tires	18 ins PAX run-flat system
Length	4280 mm (168.5 ins)
Width	1760 mm (69.3 ins)
Height	1800 mm (70.9 ins)
Wheelbase	2800 mm (110.2 ins)
Track front/rear	1545/1545 mm (60.8/60.8 ins)

Kino is a compact minivan designed to demonstrate Nissan's vision for six-seater family cars. This concept takes its design cues from the conceptual Nissan Fusion shown at the Paris auto show in 2000, notably with the gracefully arched curve of the car's nose section, but its door layout is newly asymmetrical.

The driver's side has three doors, the rearmost being rear-hinged to give better access for the third-row occupants. The passenger side gets two doors, making for a large opening, and helped by the fact that the rear door slides, as well as by the absence of a middle door pillar. The sliding door is possible thanks to the upright and fairly planar side of the Kino.

The interior is designed to be light and airy, courtesy of generous side windows, and a glass roof curving back as far as the third row of seats. The rear seats compromise luggage capacity if they are not removed. The minimalist interior can be arranged into a "living room," so occupants can sit facing each other while they watch monitors displaying video games, the Internet or TV broadcasts.

Nissan has been here before. Previously it was the Serena—dreadful from both a design and a driving-dynamics viewpoint—representing the brand in the mini-MPV sphere. The Kino seems a likely Nissan production model once some changes have been made. For instance, the lack of a conventional door pillar on the passenger side is fine for a concept model, but the missing structural rigidity caused by its absence would be unlikely to meet Nissan's crash-safety requirements—something its designers will have to cope with.

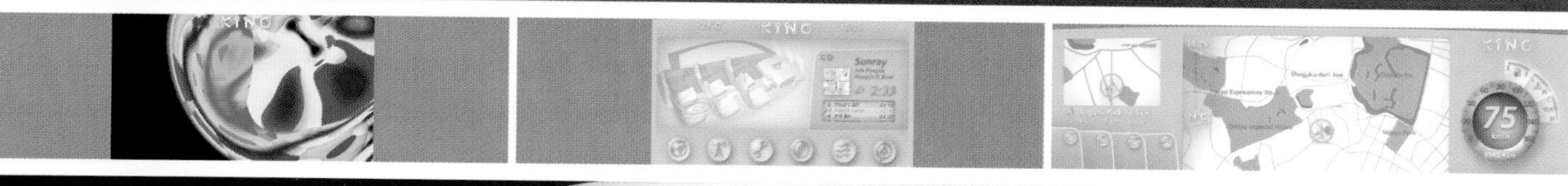

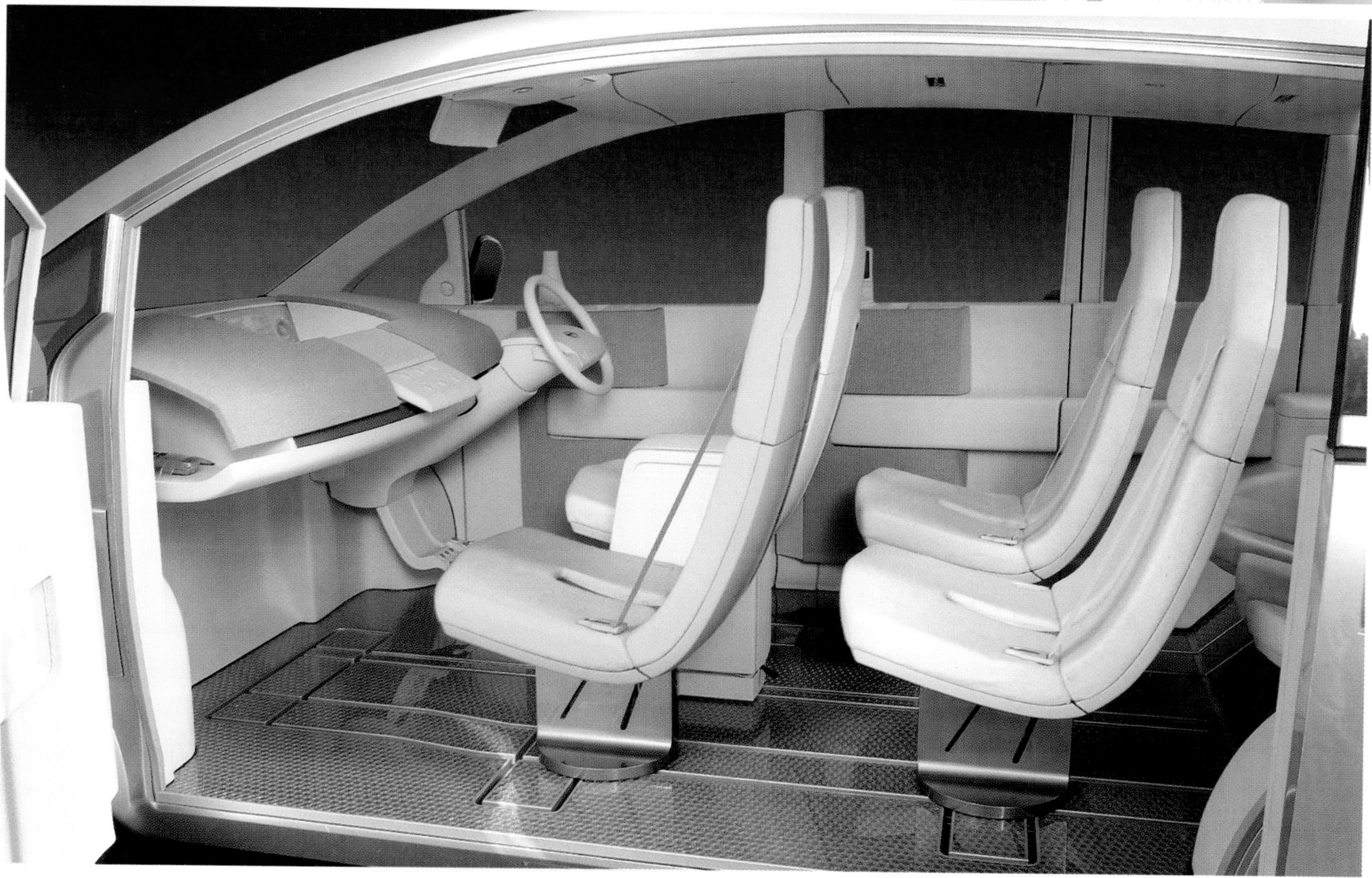

Nissan mm.e

Design	Christopher Reitz and Satori Tai
Gearbox	4-speed automatic
Front suspension	MacPherson strut
Rear suspension	Torsion beam
Front tires	175/60R15
Rear tires	175/60R15
Length	3695 mm (145.5 ins)
Width	1660 mm (65.4 ins)
Height	1525 mm (60 ins)
Wheelbase	2430 mm (95.7 ins)

Nissan calls it the mm.e, but this friendly-faced B-sector concept is really the next-generation Micra in all but a few details, unveiled at the Frankfurt Motor Show in September 2001. Within two years a slightly tamer version of the mm.e will replace the current Micra (called the March in Japan). By then the latter will have enjoyed an astonishing ten-year run as Nissan's dependable answer to the Ford Fiesta and Volkswagen Polo.

It's easy to forget that the outgoing Micra was hailed as something of a design revolution in 1992, with its tall stance, and attractively curvaceous contours. Now Nissan has redefined its entry in this sector. The mm.e's rounded appearance comprises two arches: one appearing to extend over the whole car when viewed from the side, and another that runs from the fenders, and extends across the doors at waist level.

Defining the front of the car are high-mounted, elliptical headlamps (based on LED technology), pepperpot grilles, slatted bumper intakes, and a bull-nosed hood. At the rear, the curved tailgate features a roof spoiler, plus a large rear bumper, and LED-based rear lamps.

For the interior of the car, Nissan's philosophy is to use materials and colors that create a feeling of freshness and high quality. The instrument panel is shaped like a soccer ball, with a multifunctional display. A new entry system has been designed specifically for the mm.e. Its driver carries a control unit, and locks the doors by simply pushing the corresponding button on the exterior door handle. There is no need to take out and insert a key, or to press a button on a remote, keyless entry fob.

mm

mm

NISSAN

Nissan Moco

Design	Ichi Ando and Mr Kinimoto
Gearbox	4-speed automatic
Front suspension	MacPherson strut
Rear suspension	Trailing link
Front tires	165/55R14
Rear tires	165/55R14
Length	3395 mm (133.7 ins)
Width	1475 mm (58.1 ins)
Height	1590 mm (62.6 ins)
Wheelbase	2360 mm (92.9 ins)

Setting to work with nothing more than an idea and a computer on which to translate it is one thing, but sometimes it seems concept cars are produced with only a vague notion of who the end user is meant to be. Yet Nissan is gender-specific about the Moco: this is a car for the married woman in her twenties or thirties. It is not ready for the showroom right now, but Nissan has a similarly clear idea of the source of a production Moco: if it gets the green light for the real world, Suzuki will build it for Nissan under contract.

All this is fine, but the Nissan Moco is less visually appealing than several other small cars. The Moco would be Nissan's first proper foray into city cars, a sector in which potential supplier Suzuki is a past master. Maybe those young wives will be tempted more by the brand's excellent quality, and reliability record—not to mention bargain pricing—than by the car's looks.

The exterior of the Moco is pretty much a mono-form design, with a steeply raked windshield that gives it a more forward driving position. For its size the Moco is quite tall. Even with its upright proportions it looks neat, with its simple panel forms, large headlamps, and slatted front bumper. The rear end is reminiscent of other small cars, such as the Fiat Punto, with the tailgate cut into the bumper to help when loading. The rear seats are designed to slide fore and aft for either optimum legroom or extra trunk capacity. The Moco's aesthetically unharmonious interior mixes too many differently colored and textured materials. The feeling among car designers is that Nissan needs to come up with something more attractive than this if the hearts of those twenty-something women are to be beguiled.

MOCO

MOCO

Nissan Nails

Design	Koji Nagano and Taro Ueda
Gearbox	4-speed automatic
Front suspension	Double wishbone
Rear suspension	Double wishbone
Front tires	195/60R18
Rear tires	195/60R18
Length	4250 mm (167.3 ins)
Width	1720 mm (67.7 ins)
Height	1470 mm (57.9 ins)
Wheelbase	2580 mm (101.6 ins)
Track front/rear	1500/1500 mm (59.1/59.1 ins)

A wacky oddball. That is the only way to categorize the Nails, an extraordinary design for a macho-looking pickup with a distinctly toy-like quality. The cabin is angular and aggressive, and the pickup flatbed seems just perfect for transporting off-road motorcycles or surfboards.

Nissan has tried to capture something of the strength and flexibility of the materials used in garments such as wetsuits. Extensive use of rubber on both exterior and interior surfaces gives the Nails an uncommonly resilient character. Nissan says the body panels are coated with a treatment that resists scratches or dents, making the (hard as) Nails ideal for tough, energetic misuse.

The outrageous, blue exterior panels appear to be simply bolted to a bumper-car chassis, which extends the entire length of the car. An A-frame is mounted above to form the cabin. Inside, a single seat—designed as if it were constructed from the teeth of two interlocking combs, if you can imagine that—can convert from a two-person bench to a single driver's bucket. The steering wheel, also rubber, echoes the wheels with its segmented, dished design.

There's nothing elegant about the Nails, unveiled at the Tokyo Motor Show in October 2001. It certainly makes a strong design statement, and explores the use of rubber in a street pickup. How any of the ideas it incorporates could be used in Nissan's future products is unclear, but at least it is fun, and likely to appeal to the active and anarchic snow/surf/skateboard generation of the new millennium.

0500Kg

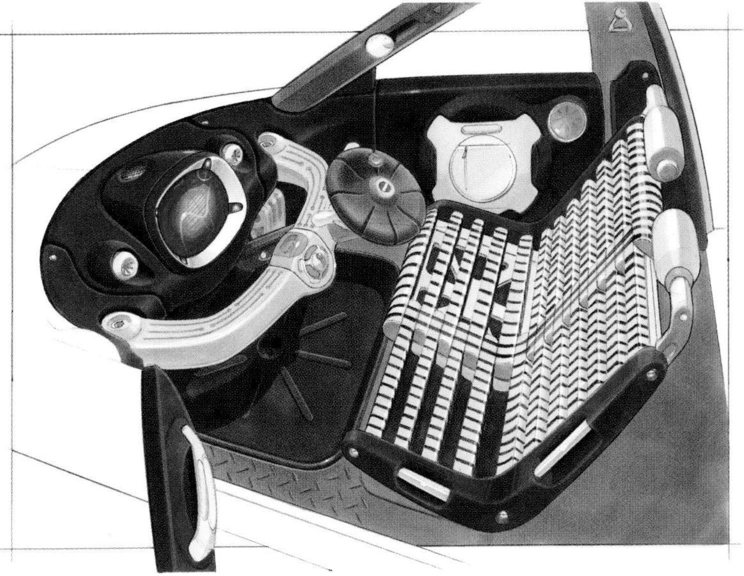

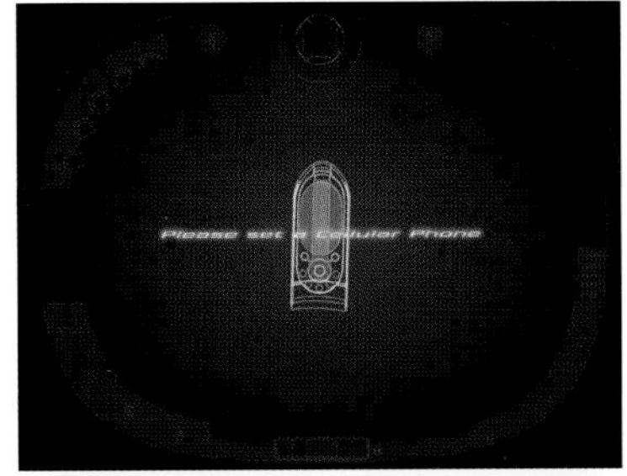
Please set a Cellular Phone

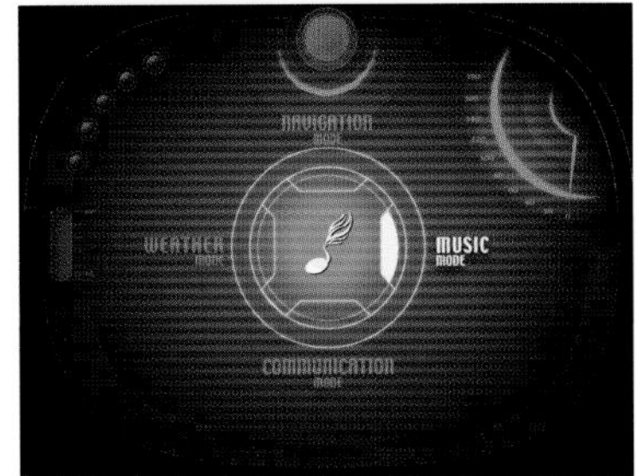
NAVIGATION
WEATHER
MUSIC
MODE
COMMUNICATION

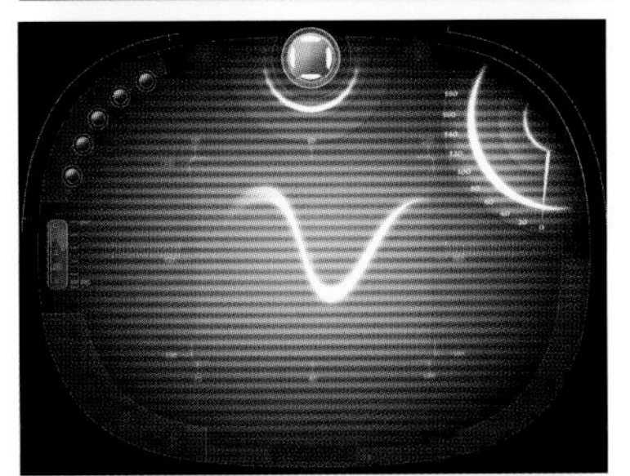

Nissan Quest

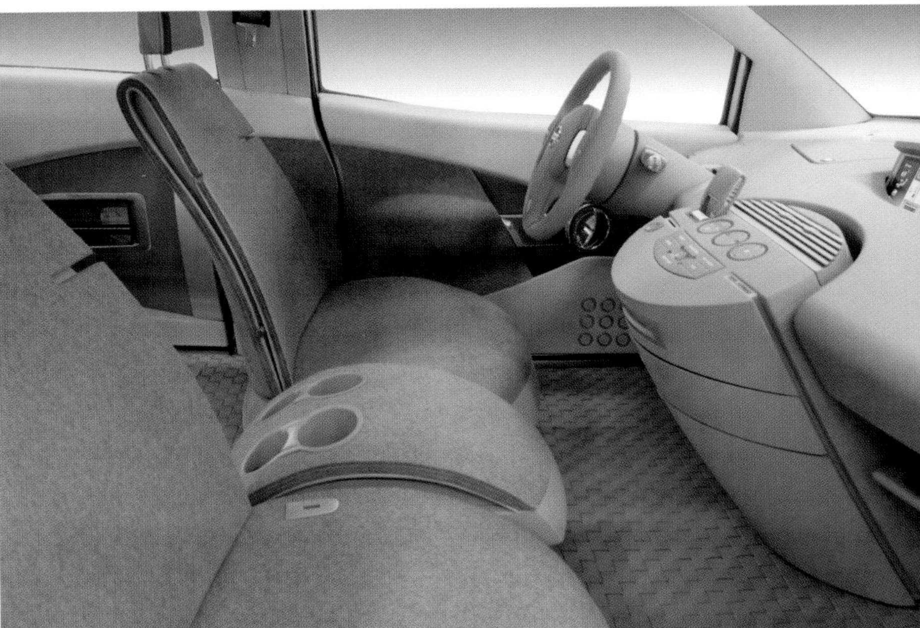

Design	Tom Semple
Engine	3.3 V6
Power	127 kW (170 bhp) @ 4800 rpm
Torque	271 Nm (200 lb ft) @ 2800 rpm
Gearbox	4-speed automatic
Installation	Front-engine/front-wheel drive
Front suspension	Independent strut suspension
Rear suspension	Beam axle with leaf springs
Brakes front/rear	Discs/drums, ABS, rear load-sensing proportional valve
Front tires	225/60R16
Rear tires	225/60R16
Length	4943 mm (194.6 ins)
Width	1902 mm (74.9 ins)
Height	1709 mm (67.3 ins) (with roof rack)
Wheelbase	2850 mm (112.2 ins)
Track front/rear	1610/1610 mm (63.4/63.4 ins)
Curb weight	1840 kg (4057 lb)
Fuel consumption	12.3 ltr/100 km (19.1 mpg)

Nissan will go it alone with the replacement for its US-market-only Quest minivan, having dissolved a joint manufacturing and design agreement with Ford for the previous model. Thus this concept shows the freedom in Nissan's future thinking as it prepares to take on US-market giants such as the Chrysler Voyager and the Honda Odyssey.

The main theme is an attempt to demonstrate that style and functionality can go together in minivans, vehicles that are usually characterized by bland design. An unusual arched theme runs through the body form, with visual breaks helping to eliminate the traditional one-box, one-volume appearance.

Two key elements lead the Quest concept's departure from traditional exterior styling: bold, architectural forms, with a flowing, arched belt line, and the extensive use of glass, such as the wraparound glasshouse with hidden pillars, and full-length glazed roof panel. This glazed design theme tries to take full advantage of the minivan's core strength of cabin roominess by increasing the amount of light that enters through the roof and windows. Architects have long grappled with this problem. The Quest borrows some building solutions: for example, the side windows with their milled edges reinforce the architectural feel and flow of the body, especially when viewed in profile.

One practical feature is the four exterior doors that open wide for easy access. The tailgate features a novel "bi-fold" design, able to operate in a confined space such as a garage or car park.

The exterior's arc theme is repeated inside on the instrument panel, door openings, and seats. Other features include a camera to monitor an infant in a rear-facing child seat, and a camera to replace the rear-view mirror (also a feature on the Acura RD-X; see p. 20).

This is a noteworthy attempt by Nissan to redefine the minivan. Perhaps the Quest represents the future exemplar for this currently dull market sector.

Nissan Yanya

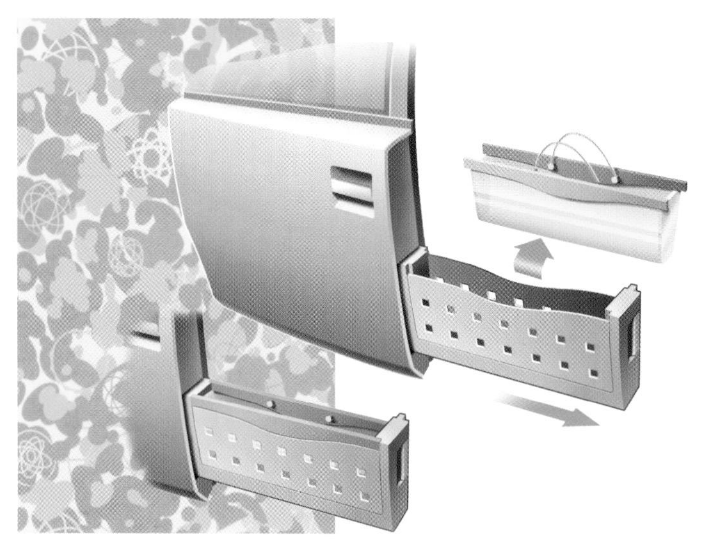

Design	Masato Inoue
Installation	Front-engine/front-wheel drive

The innovative Yanya B-sector, sports-utility-vehicle concept is Nissan's attempt to define new limits of flexibility and technology for the next generation of sophisticated and active young people.

Stylish exterior-design cues hint at the Yanya's 4×4 credentials, while contrasting with the soft-touch, durable interior materials. However, the Yanya is more than an SUV. Through its adaptable interior and roof section it transforms into a sports-utility truck. But there's more: as well as serving as a passenger car or load lugger, the Yanya's versatility means it can morph into a mobile living room, complete with its own table.

Neat detailing contributes to the freshness of the Yanya's exterior. Its grille is a pattern of small squares, reminiscent of traditional Japanese decor. This motif is echoed in the headlamps and tail lamps, and also in the custom-cut tread pattern of its run-flat tires.

Rectangular decorative elements adorn the broad air intake in the front bumper and wheel-arch extensions. Several of the rectangular indents in the titanium-colored stripes on the upper bodywork disguise mounting-points for roof-racks, additional driving lamps, and other exterior accessories.

The Yanya's social character is not restricted to its interior. When the party's going on outside, a large and stylish entertainment box, complete with speakers and internal power supply, can be removed from the center of the dash and carried out.

Nissan is targeting buyers who are young and sociable, but above all individual. They have, it says, an inherent sense of contemporary style, and are forward-thinking.

yanya

Nissan Z

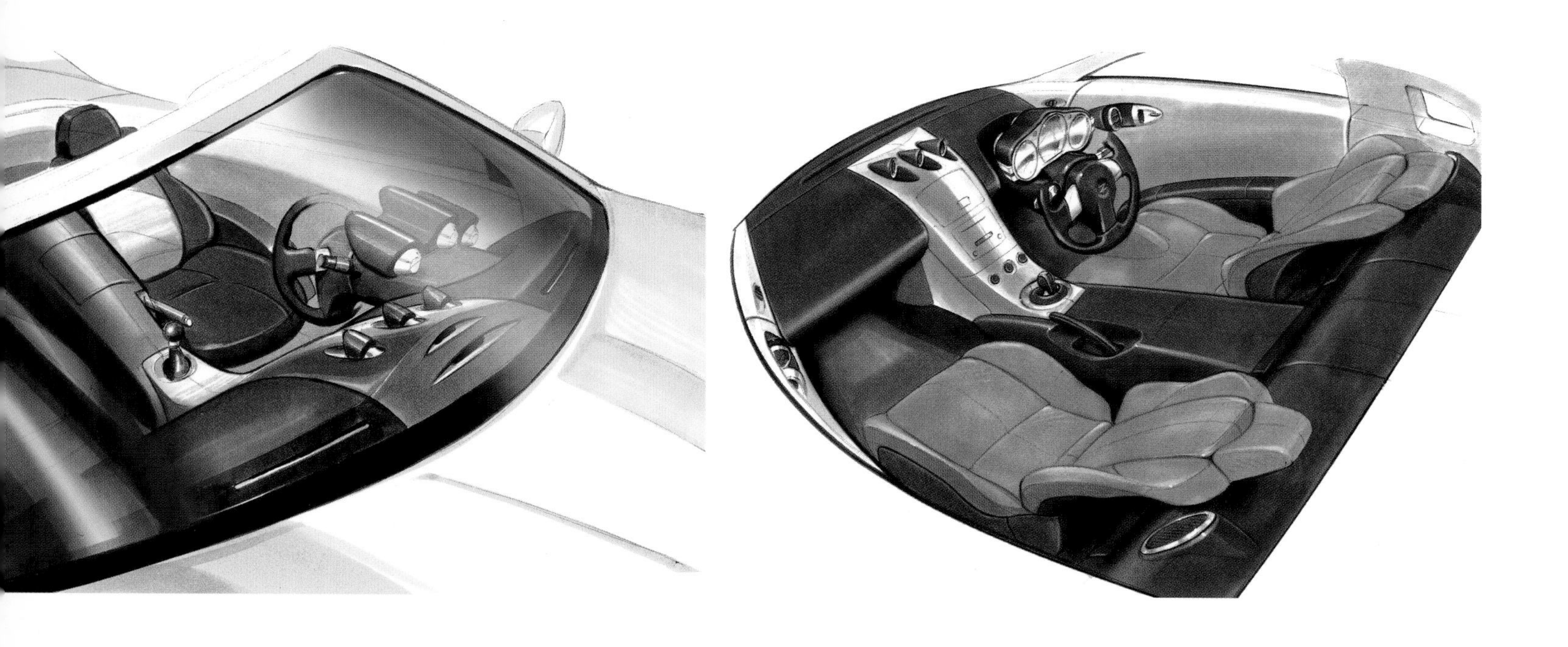

Design	Diane Allen
Engine	3.5 V6
Power	209 kW (280 bhp) @ rpm
Torque	353 Nm (260 lb ft) @ rpm
Gearbox	6-speed manual/5-speed automatic
Installation	Front-engine/rear-wheel drive
Front suspension	Independent multi-link, stabilizer bar
Rear suspension	Independent multi-link, stabilizer bar
Brakes front/rear	Discs/discs, ABS, Brake Assist, EBD, TCS, VDC
Front tires	225/50WR17
Rear tires	235/50WR17
Length	4310 mm (169.7 ins)
Width	1815 mm (71.5 ins)
Height	1310 mm (51.6 ins)
Wheelbase	2650 mm (104.3 ins)
Track front/rear	1535/1535 mm (60.4/60.4 ins)

The new Z from Nissan, unveiled at the Tokyo Motor Show in October 2001 in production-ready form, was originally shown as a concept at the Detroit auto show in January that year. It has changed only in minimal detail from the earlier showpiece. The headlamps have sharper corners, the wheels have a new center-cap design, the exhausts are round instead of rectangular, and the front-fender "Z" emblem is now contained within a chrome ring.

The Z steps into the shoes of a sports-car institution. The original, the 240Z, made its debut in 1970. It helped transform Nissan's mundane image with its looks, power, and prowess on the world rally stage. Design cues inherited from that classic model include the long nose, the triangular cabin profile, and the arch-shaped roof extending into the hatchback opening.

Contrasting with the soft-looking body are striking geometric features such as the headlamps, door mirrors, and door handles, and, at the rear, the lamps and rear screen. The body's center of gravity is in the middle, and the wheels are at the four corners, both of which help give the Z an agile stance.

The cabin is a snug fit for driver and passenger, with a high, central transmission tunnel separating its seats. Front legroom is tapered not only for a sporty feel, but also to accommodate the chunky front wheels, and the suspension tunnel. On the dash, the three main instruments tilt with the steering column, so that they're clearly visible regardless of where the driver's seat is positioned. Door trims are composed of soft, emotive surfaces, while the door handles are contrastingly sharp—a common design approach today.

It seems that the Z cannot fail to be a hit, so carefully has Nissan assimilated its sports-car heritage into a thoroughly modern package. The car pushes the right enthusiast buttons, and Nissan's reputation for quality only adds to its attractiveness.

Opel Concept M

Design	Hans Seer
Engine	1.6 turbo, running on natural gas
Power	110 kW (150 bhp)
Torque	205 Nm (151 lb ft) @ 1980 rpm
Gearbox	5-speed Easytronic automated manual
Installation	Front-engine/front-wheel drive
Length	4050 mm (159.4 ins)
Width	1780 mm (70.1 ins)
Height	1620 mm (63.8 ins)
Wheelbase	2630 mm (103.5 ins)
0–100 km/h (62 mph)	9.8 sec.
Top speed	202 km/h (125 mph)
CO_2 emissions	145 g/km

This dynamically styled Opel design study gives an insight into a new type of "monocab" passenger van.

Despite its compact exterior dimensions, close to those of the Mercedes-Benz A-Class, the Concept M has a spacious interior for four passengers. Moreover, it offers considerable potential for innovative interior solutions in the future.

The Concept M's exterior borrows from the styling of the Signum2 (see p. 198), while the striking Plexiglas headlamps, and the chrome strip across the tailgate, are clearly derived from the new Opel/Vauxhall Vectra (see p. 200). The conspicuous, full-width brake light along the top of the rear window displays, in three levels of brightness, how hard the brakes have been applied.

The overhead console in the cabin houses reading lamps, and an "AutoVision" entertainment system with a DVD player. In a touch again borrowed from the Signum2, the variable center console extends into multi-purpose aluminum tracks between driver and passenger, with holders for cups, utensils, or a laptop computer.

The Concept M instrument panel houses a fully configurable 228 mm (9 ins) screen that provides the driver with vehicle speed and engine-speed information, displayed on two circular, easy-to-read gauges. The navigation system indicates a planned turn with a large arrow icon that points in the chosen direction. This futuristic "Info Center" display can be adjusted to suit the driver's preferences by, for example, choosing a different background color.

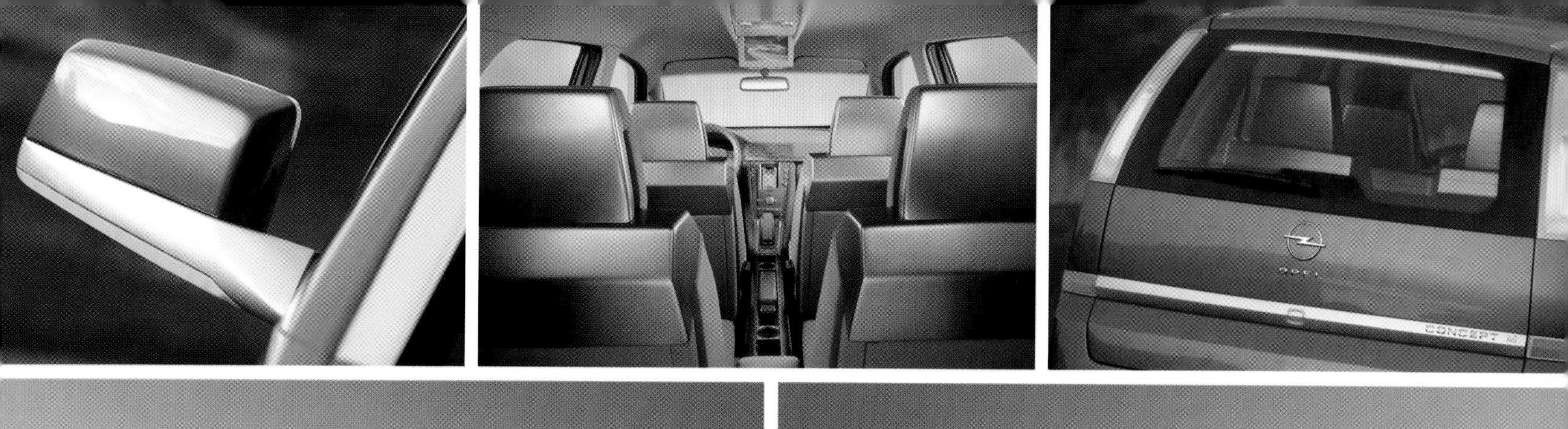
CONCEPT

GG BJ 941

GG BJ 941

Opel Frogster

Design	Stefan Arndt
Engine	1.2 in-line 4
Power	43 kW (58 bhp)
Gearbox	Easytronic automated manual
Front tires	175/55R15
Rear tires	175/55R15
Length	3715 mm (146.3 ins)
Width	1680 mm (66.1 ins)
Height	1530 mm (60.2 ins)

Opel offers the Agila city car, and the Frogster is a design study for a possible convertible version of that model. This concept is an interesting one, because the Agila has a tall, upright body, intended to squeeze as much passenger space into as small a road footprint as possible. But such proportions do not match the classic, low-slung, sporty look of a traditional convertible, so the Frogster presents a highly unusual shape. And since the Agila is itself a low-cost car, the Frogster could carry a competitive price tag attractive to younger buyers. Thus its styling reflects its intended appeal to what Opel dubs the "Playstation generation."

Technology is also part of the Frogster's design. A mini-computer buried in the dash is remotely accessed from a Personal Digital Assistant (PDA). This link is claimed to work even when the PDA is in the driver's pocket, and is programmed with various functions, including one that replaces the conventional key. When housed in the center console, the PDA also becomes a display for the radio, navigation, and climate-control systems.

Weather protection is novel too. In place of the usual folding soft-top, an electrically powered roller cover provides shelter. The roof can be operated from the PDA. Just in case the weather gains the upper hand, the interior is weather-proofed. Like surfers' body suits, the seats are covered in Neoprene, the same material used for the cellphone holder.

Another practical detail is the individually folding seats that can switch the Frogster from a one- or two-seat roadster into a three- or four-seat convertible, and even a pickup.

And that name? Back in 1924, Opel enjoyed success with a small, open car (with 12 hp). Thanks to its green paint, it was nicknamed the Laubfrosch or Tree Frog.

FROGSTER

Opel Signum2

Design	Hans Seer
Engine	4.3 V8
Power	220 kW (300 bhp)
Torque	400 Nm (295 lb ft)
Gearbox	5-speed automatic
Installation	Front-engine/front-wheel drive
Brakes front/rear	Discs/discs
Length	4640 mm (182.7 ins)
Width	1790 mm (70.5 ins)
Height	1460 mm (57.5 ins)
Wheelbase	2830 mm (111.4 ins)
Top speed	250 km/h (155 mph)
Fuel consumption	9.8 ltr/100 km (23.9 mpg)

Opel's Signum2 gives the impression of great strength. Its rising belt line has a sharp edge running along its length that continues through to the rear lamps. The exterior is generally unfussy, with large, undecorated surfaces. At the rear of the car the tailgate's upper half is fully glazed, while its lower half incorporates a feature that echoes the sharp edge along its length. A glass panorama-roof ensures a spacious, airy feel to the interior which, Opel claims, can accommodate four occupants in business-class comfort.

A new feature of the Signum2 is the rear seats. Their backs fold upward to the height of the belt line to create a level surface, beneath which luggage can be stowed, out of sight, and well secured. When the front door handle is pulled, the front seats swivel automatically toward the doorways, while the steering "wheel"—whose odd shape makes it easier to see the instruments but would meet with market resistance—lowers itself into the instrument panel. When the driver rotates his seat forward again, the wheel re-emerges at the push of a button. The front-seat passenger faces an "infotainment" screen. When not in use, the screen folds flat, and disappears into the instrument panel above the glove compartment. Passengers in the rear can watch DVDs through specially-designed video glasses.

Opel is in trouble in Europe, making both heavy losses and sometimes forgettable products. Its first attempt at a new direction for executive cars, also called Signum, took its bow in Frankfurt in 1997. The Signum2 builds on the earlier concept for General Motors' European arm. Its ideas could shape future Opel (and Vauxhall) models. The new Vectra adopts much of its obvious design solidity; it is a crucial car on which a lot is riding.

DVD

Opel/Vauxhall Vectra

Design	Michael Pickstone and Malcolm Ward
Engine	3.2 V6 (1.8 and 2.2 in-line 4, and 2.0 and 2.2 in-line 4 turbo-diesel, also offered)
Power	155 kW (211 bhp) @ 6200 rpm
Torque	300 Nm (221 lb ft) @ 4000 rpm
Gearbox	5-speed manual/5-speed automatic
Installation	Front-engine/front-wheel drive
Front suspension	MacPherson strut
Rear suspension	Multi-link
Brakes front/rear	Discs/discs
Front tires	215/50R17
Rear tires	215/50R17
Length	4596 mm (180.9 ins)
Width	1798 mm (70.8 ins)
Height	1460 mm (57.5 ins)
Wheelbase	2700 mm (106.3 ins)
Track front/rear	1535/1525 mm (60.4/60 ins)
Curb weight	1395 kg (3076 lb)
0–100 km/h (62 mph)	7.5 sec.
Top speed	248 km/h (154 mph)
Fuel consumption	10.1 ltr/100 km (23.2 mpg)
CO_2 emissions	243 g/km

Opel has a long history in the European mid-size category; since 1970 the company has sold almost 9 million cars in this sector worldwide. The most recent Vectra, however, was a big disappointment—a mediocre car given a well-deserved drubbing by critics. Its poor sales were one reason given for the closing of Vauxhall's British plant at Luton, north of London, in 2002 after nearly one hundred years of car production.

The new Vectra, its replacement, made its debut in spring 2002, beginning with the four-door sedan. A few months later the Vectra GTS joined it. Sporting unique coupe-like lines, four doors, and a tailgate, it is aimed at younger buyers who want a racier car.

The third-generation Vectra is the first volume model to adopt Opel's new design style. The Vectra design team focussed on two key themes: tension and motion.

The trapezoid-shaped radiator grille at the front is integrated into the hood, and a distinctive chrome bar has a large Opel (or Vauxhall) emblem in the middle. Large tail lights that wrap around the corners dominate the view from the rear, and culminate in the wide shoulder line. A distinctive stainless-steel strip at the bottom of the trunk lid links the lights visually, and gives an impression of width.

Anthracite is the color chosen by the interior designers for the upper part of the instrument panel, because it helps prevent distracting reflections on the windshield. This elegant, dark, matt finish continues in the upper door panels, and in the rear storage area.

The Vectra GTS looks sportier, and more aggressive, than its elegant sedan counterpart. Its distinctive appearance comes across through a characteristic nose treatment with darkened headlamps, a different bumper with larger air intake, and round fog lights instead of rectangular ones.

VECTRA

Peugeot 807

Design	Nicolas Brissoneau
Gearbox	Manual or automatic
Installation	Front-engine/front-wheel drive
Front suspension	MacPherson strut
Rear suspension	Deformable beam
Brakes front/rear	Discs/discs, ABS, EBD, ESP
Length	4720 mm (185.8 ins)
Width	1854 mm (73 ins)
Height	1856 mm (73.1 ins)
Wheelbase	2823 mm (111.1 ins)
Track front/rear	1570/1548 mm (61.8/60.9 ins)

Like Renault's Espace concept, the 807 full-size MPV enhances the wellbeing of its occupants with extensive storage capacity, a sense of spaciousness, and a generally relaxed environment. This it achieves by having a large "monospace" cabin, and a window-forward design. The 807 has a panoramic windshield (its surface area is 1.95 m^2/2.2 sq yd)—the largest in any car in standard production today.

Peugeot unveiled the 807 at the Geneva International Motor Show in 2002. It is the brand's new-generation, top-of-the-range people-carrier, replacing the 806, and sharing much with the Fiat, Citroën, and Lancia models built at the same French plant.

The front of the 807, has the familiarly styled headlamps and grilles featured on the Peugeot 206 and 307. The sliding rear doors can be motorized, with an obstacle-detection sensor, and operated by remote control. Door handles are painted a satin-aluminum finish for a quality feel.

Inside the 807 a conventional panel is replaced by an architecturally inspired "arch" that rests on a central pillar. The instruments and storage areas sit in this unit, and are covered with either leather or cloth.

Available as an option is a color screen, incorporated in the roof, that unfolds along the centerline of the vehicle, and can be seen by all the rear passengers. The screen can be connected to a games console, DVD player, camcorder, or computer and, for peaceful driving, can deliver audio through infra-red headphones.

252 NSJ 75

Peugeot RC

Design	Nicolas Brissoneau
Engine	2.2 in-line 4 diesel (2.0 in-line 4 also offered)
Power	129 kW (173 bhp)
Torque	400 Nm (295 lb ft)
Gearbox	6-speed manual
Installation	Rear-engine/rear-wheel drive
Front suspension	Double wishbone
Rear suspension	Double wishbone
Brakes front/rear	Discs/discs
Front tires	245/45R18
Rear tires	245/45R18
Length	4300 mm (169.3 ins)
Width	1800 mm (70.9 ins)
Height	1150 mm (45.3 ins)
Wheelbase	2800 mm (110.2 ins)
Track front/rear	1600/1500 mm (63/59.1 ins)
Curb weight	950 kg (2095 lb)
0–100 km/h (62 mph)	6 sec.
Top speed	230 km/h (143 mph)
Fuel consumption	4.9 ltr/100 km (47.9 mpg)
CO_2 emissions	132 g/km

Peugeot brought two unlikely concepts to the Geneva International Motor Show in 2002. The RCs are a pair of rear-engine sports coupes aimed at flaunting Peugeot's dynamic credentials from a design and technological standpoint. Named after the playing cards, the RC Diamonds is suitably red, and fitted with the 2.2 ltr HDi diesel engine, while the RC Spades is jet-black, and has a 2 ltr gasoline engine.

Aerodynamic management has a major influence on the exterior design of the RC. Its low height, clean profile, air outlets behind the front wheels, and diffuser at the rear combine to achieve a low-drag shape. A retractable aerofoil at the rear provides extra downforce at speed.

Several advanced technologies are used in the RC. In racing-car tradition, the structure is made from carbon pre-impregnated directly on to honeycomb panels, formed and baked in an autoclave. This technique is currently only feasible for low-volume production models, so the RCs are unlikely to come to market in their current form. Double-wishbone suspension, and ceramic brake discs, ensure excellent ride and handling, giving good wheel control and low unsprung mass.

The interior styling is sporty, using materials combining red and black leather and cloth for the body-hugging seats. The leather-upholstered dash panel houses a chronograph-style instrument panel with digital and analog readouts.

Access to the rear seats of this two-plus-two is gained via fantastic, beetle-winged doors, and sliding front seats.

2.08
2.08
2.08

Pininfarina Ford Start

Design	Pininfarina and Chris Bird
Engine	2.0 in-line 4 turbo
Power	164 kW (220 bhp)
Front tires	255/40ZR19
Rear tires	255/40ZR19
Length	4200 mm (165.4 ins)
Width	1830 mm (72 ins)
Height	1300 mm (51.2 ins)
Wheelbase	2515 mm (99 ins)
Track front/rear	1600/1585 mm (63/62.4 ins)
Top speed	242 km/h (150 mph)

Pininfarina is one of the truly great names in car design. It has been linked to everything from Peugeot to Ferrari, but never to Ford—until now.

First, the bad news: the Ford Start, which Pininfarina has designed for Ford of Europe, will never be produced. According to the Italian company, it is purely a concept exercise, despite the fact that Pininfarina has deliberately echoed Ford's design language with a look based on angular lines and emphatic shapes, aiming for a design that combines current fashions with the architecture that underlies the dynamic Ford style.

Now the good news: the result had a positive reception throughout the industry when it was unveiled at the Frankfurt Motor Show in September 2001. It is a two-plus-two, high-performance sports car defined by its balanced masses, its long, sweeping line from roof to rear window, its large wheelarches, and six-spoke wheels. The clean-cut shape is—as you might expect from Pininfarina—deliberately simple and elegant, although it is uncanny how similar the car is to Nissan's new Z (see p. 192). A key Start feature is its advanced roof system, which can change the coupe into a convertible at the push of a button. The roof folds away into a storage area at the rear.

The interior of this one-off show car is handcrafted with a typically Italian sumptuousness. Its surfaces are simple to the point of austerity, while the tub-shaped front seats combine clean lines with body-hugging forms.

Finally, the really good news: the Pininfarina–Ford partnership is going from strength to strength. The Start has turned out to be just that. A further result of this creative relationship is the Pininfarina-designed Ford StreetKa, an attractive Ka-based sports convertible being engineered for production right now.

Pontiac Solstice

Design	Franz von Holzhausen
Engine	2.2 in-line 4 supercharged
Power	179 kW (240 bhp)
Torque	305 Nm (225 lb ft)
Gearbox	6-speed manual
Installation	Front-engine/rear-wheel drive
Front suspension	Independent with adjustable struts
Rear suspension	Independent with adjustable struts
Brakes front/rear	Discs/discs
Front tires	245/35R19
Rear tires	255/35R20
Length	3904 mm (153.7 ins)
Width	1805 mm (71.1 ins)
Height	1143 mm (45 ins)
Wheelbase	2415 mm (95.1 ins)
Track front/rear	1532/1537 mm (60.3/60.5 ins)
Curb weight	1318 kg (2900 lb)

Of the many companies that have tried, so far only Mazda—with its neat MX-5 Miata—has succeeded in producing an affordable front-engine two-seat sports car that revives the fun-to-drive feeling of the classic British and Italian roadsters of the 1960s.

The Pontiac Solstice, launched at the Detroit auto show in January 2002, is a concept for just such a car. It is General Motors' bid to muscle in on the sports-car scene, and, at the same time, to inject much-needed sparkle into the corporation's line-up of products.

The Solstice is very much the inspiration of GM's new product boss, Bob Lutz. It reflects his reputation as a "car guy" capable of whipping up excitement both within a company, as he did at Chrysler, and among the buying public.

Conceived and built in a matter of three months, the Solstice is in fact two cars—a convertible roadster, and a curvy fastback coupe. Both share the same Opel/Vauxhall Astra platform, and a supercharged 2.2 ltr engine, also drawn from the Astra. Crucially for the appeal of a sports car, however, the 179 kW (240 bhp) engine drives the rear wheels, promising a thrilling, as well as a stylish, drive.

The Solstice roadster, especially, has that cheeky "get in and drive" look that is hard to resist. It is short and solid, yet also rounded and flowing, cleverly incorporating design cues from a whole variety of classic sports cars. The interior is as devoid of surplus detail as the exterior, reinforcing the feeling that this is a car designed for enthusiasts rather than show-offs.

Many at Detroit pointed to the Solstice duo as the most tempting concepts of the show. No doubt Bob Lutz and the General Motors bosses will have taken note of the uncharacteristic buzz echoing through the Pontiac stand during the exhibition.

Pontiac Vibe

Design	John Mack
Engine	1.8 in-line 4
Power	134 kW (180 bhp) @ 7600 rpm
Torque	176 Nm (130 lb ft) @ 6800 rpm
Gearbox	6-speed manual
Installation	Front-engine/front- or 4-wheel drive
Front suspension	MacPherson strut
Rear suspension	Torsion beam with trailing link
Brakes front/rear	Discs/discs, ABS
Front tires	215/50ZR17
Rear tires	215/50ZR17
Length	4365 mm (171.9 ins)
Width	1775 mm (69.9 ins)
Height	1590 mm (62.6 ins)
Wheelbase	2600 mm (102.4 ins)
Track front/rear	1505/1485 mm (59.3/58.5 ins)
Curb weight	1273 kg (2807 lb)
0–100 km/h (62 mph)	8 sec.
Fuel consumption	10.1 ltr/100 km (23.2 mpg)

The most significant aspect of the 2003 Pontiac Vibe is that, underneath, it is a Toyota Corolla. It is closer still to the Toyota Matrix, a lifestyle-oriented Toyota station wagon also built on the Corolla platform. The pair not only share just their engines, chassis, electrics, and interior, but are also built in the same California factory.

As its chosen name suggests, Pontiac is aiming the Vibe at a young and hip audience. The model is designed to be the snowboard generation's take on the small station wagon, with a hefty dose of high-revving power from Toyota's variable-valve-timing engine thrown in for good measure.

Despite the bruising it received for the downright ugly Aztek SUV, Pontiac has not been deterred from an adventurous look for the Vibe. Visually it manages to be tough, sporty, and sharp at the same time. The design recalls some traditional Pontiac styling cues, such as the twin-port grille, the cat-eye headlamps, and the recessed fog lamps. The racy, overall impression is softened by plastic lower-body cladding, which hints at potential off-road adventures. Four-wheel drive is indeed available, although only with the less powerful of the two 1.8 ltr engine choices.

To suit the lifestyle of its target owner group, the Vibe's interior had to combine essential electronic gadgetry—the stereo system produces over 200 watts—with adequate space, and a versatile cargo bay for hauling sports equipment. The load area is well thought out, with numerous tie-down cleats, folding seats, and a waterproof space for wet gear in the spare-tire well. The claim is that this compact wagon, with the performance of a sports coupe (the engine is shared with the potent Celica GT), can carry two mountain bikes in the back. Not bad for under $19,000 argues Pontiac.

Vibe

Range Rover

Design	Geoff Upex
Engine	4.4 V8 (3.0 in-line 6 turbo-diesel also offered)
Power	210 kW (282 bhp) @ 5400 rpm
Torque	440 Nm (325 lb ft) @ 3600 rpm
Gearbox	5-speed Steptronic automatic
Installation	Front-engine/4-wheel drive
Front suspension	Electronic air suspension with variable ride height
Rear suspension	Electronic air suspension with variable ride height
Brakes front/rear	Discs/discs
Front tires	205/70R15
Rear tires	205/70R15
Length	4950 mm (194.9 ins)
Width	2009 mm (79.1 ins)
Height	1863 mm (73.3 ins)
Wheelbase	2880 mm (113.4 ins)
Track front/rear	1629/1626 mm (64.1/64 ins)
Curb weight	2570 kg (5667 lb)
0–100 km/h (62 mph)	9.2 sec.
Top speed	208 km/h (129 mph)
Fuel consumption	16.2 ltr/100 km (14.5 mpg)
CO_2 emissions	389 g/km

The new Range Rover was unveiled at the Design Museum in London. This was quite an event, not least for the fact that this is only the third new Range Rover model in thirty-two years. However, unlike the outgoing version, this is an all-new car from end to end. To be sure, it does resurrect some design touches from the original—a vehicle so fêted that it remains the only car ever exhibited at the Louvre in Paris as an example of modern sculpture.

The design team was led by Geoff Upex, Land Rover design director. His aim was to create a vehicle blending the key styling cues of the existing model with eye-catching modern "jewelry." This is achieved in a number of ways. The headlamps, now circular again just as they were in 1970, sit behind clear glass. The rear lamps are much smaller, and are body-mounted only, making the tailgate look wider, and less cluttered, and front and rear bumpers are incorporated into the body design, flush-mounted, and body-colored.

Upex's team has created a more solid form, making the new Range Rover look much more like a luxury car, and less like an off-roader. It's bigger on the outside—longer, wider, and with even greater ground clearance. The sumptuous impression is enhanced by a far more spacious cabin.

The interior takes styling and texture influences from products as diverse as audio equipment, ocean-going yachts, first-class airline seating, fine furniture, and jewelry. These concepts have been artfully combined with the classic "wood-and-leather" look instantly associated with Range Rovers.

The new Range Rover remains a highly recognizable product that does not stray far from its historical roots. Land Rover is tailoring its new model much further toward the luxury-car sector—it is, after all, more profitable than the 4x4 off-road market.

RANGE

Renault Espace

Design	Patrick le Quément

Renault has a challenge on its hands with the new Espace: how does it push back the boundaries of a design idea—the multi-purpose vehicle, or MPV—that it pioneered back in 1984?

The new Espace concept's debut was a genuine highlight of the 2002 Geneva show. It has been conceived with a strong new personality, sporting the taut lines, and distinctive edginess, that Renault embarked upon with its Vel Satis. But the real innovation, and design energy, have gone into the interior.

A key project goal was to enhance the comfort, and wellbeing, of passengers. Heating and ventilation controls are now accessible to all occupants, while three-point, seat-mounted safety belts give total autonomy from other passengers.

The "shared" dash can be seen by every rear-seat occupant. It blends seamlessly into the cabin with pure, minimalist lines. Instruments appear as luminous blue—rather like the Volkswagen Golf's—against a black background, ensuring a clear, and highly legible, display.

The pale, grayish-beige materials inside are welcoming, and restful. The seats and armrests, upholstered in pale-beige leather, and the floor mats add further warmth to the cabin. In contrast, the lower dash, and the central information console, are a glacé chestnut color.

The Espace concept makes ample use of transparency and illumination. It boasts what will be the biggest glass sun-roof on the market, covering an enormous 2.16 m^2 (2.4 sq yd) in a vast, glazed expanse. Fortunately it can be screened off with a blind on baking-hot days.

3.0dCi

Renault Talisman

Renault's Talisman, with its distinctive gull-wing doors, is another exciting concept from Patrick le Quément. This grand tourer has a rounded, sculptural form, muscular and well proportioned. The highly distinctive, arc-shaped side windows, made from a single piece of glass, add to the car's overall dynamism. Headlamps—based on fluid-optics technology—are fashioned from vertical glass strips extending into the tops of the front fenders.

Gull-wing doors have a checkered career as a design feature. They were pioneered by Mercedes-Benz in 1954; the only other production cars to feature them were the blighted 1975 Bricklin and 1982 De Lorean. Fitted to the 1980 Aston Martin Bulldog prototype, they drenched the driver when he opened them on wet days. Just like on the Talisman, they looked dramatic on the 1966 Lamborghini Marzal when they opened up the entire side of the car. The Talisman's doors, 2.45 m (8 ft) long, open electro-hydraulically to reveal a simple interior with no central pillar that might restrict access to the cabin.

The Talisman's seats are solidly fixed. The dash and pedal unit can be electrically adjusted forward or backward to give the driver a perfectly comfortable position. The advantages are, first, that rear-seat passengers always have the same decent amount of legroom, and, secondly, the cavities normally formed between the door panels and seats are closed, which is both more comfortable and safer.

Technology showcased on the Talisman includes three cameras that relay a panoramic view of the outside environment to a wide-screen display, fitted with voice-activation, located in the dash top. Also new on the Talisman is Touch Design (TD), pointing to the future philosophy of Renault interiors. Under TD, each control must suggest its function, encouraging people to reach out and use it. Le Quement aids this by combining ergonomics, handsomeness, and materials that are soft to the touch.

Talisman is a flight of fantasy. In any case Renault has its hands full with models featuring "offbeat" doors already; it had to delay the launch of its new luxury Avantime "executive coupe" several times because its two enormous and heavy doors were tricky to perfect. So don't expect to see the Talisman anywhere but at an auto show for now.

Design	Patrick le Quément
Engine	4.5 V8
Gearbox	Automatic
Installation	Front-engine/rear-wheel drive
Brakes front/rear	Discs/discs
Front tires	21 ins PAX run-flat system
Rear tires	21 ins PAX run-flat system
Length	4805 mm (189.2 ins)
Width	1950 mm (76.8 ins)
Height	1380 mm (54.3 ins)
Wheelbase	2950 mm (116.1 ins)
Curb weight	1600 kg (3528 lb)

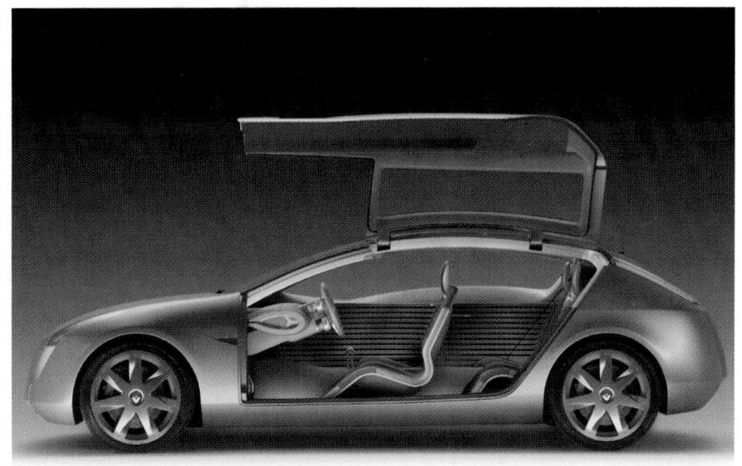

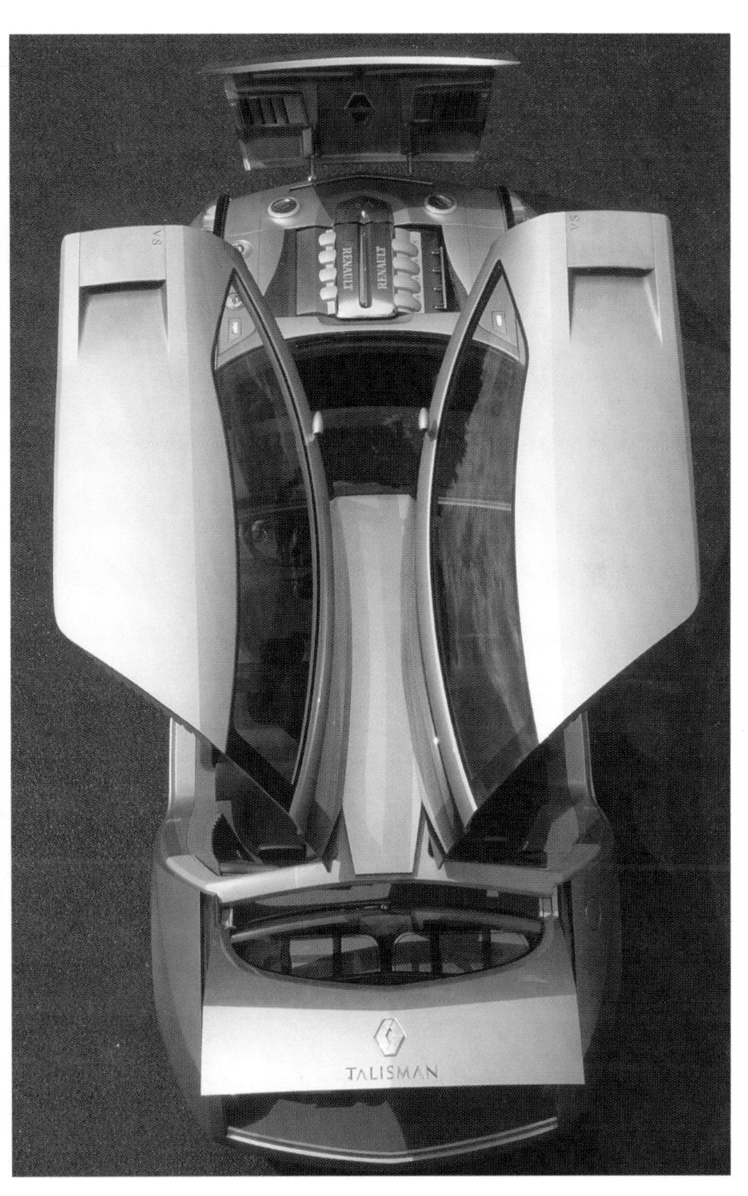
RENAULT
TALISMAN

Rinspeed Presto

Design	Thomas Clever
Engine	1.7 in-line 4 using natural gas
Power	88 kW (120 bhp) @ 4200 rpm
Torque	224 Nm (165 lb ft) @ 1600 rpm
Gearbox	5-speed automatic
Installation	Front-engine/front-wheel drive
Front suspension	MacPherson strut
Rear suspension	Semi-independent
Brakes front/rear	Discs/discs
Front tires	205/50R17
Rear tires	235/45R17
Length	2996–3742 mm (118–147.3 ins)
Width	1847 mm (72.7 ins)
Height	1375 mm (54.1 ins)
Wheelbase	1750–2496 mm (68.9–98.3 ins)
Track front/rear	1507/1580 mm (59.3/62.2 ins)
Curb weight	865 kg (1907 lb)
0–100 km/h (62 mph)	10.5 sec.
Top speed	180 km/h (112 mph)

The transforming Rinspeed Presto converts itself from two-seat roadster, less than 3 m (9.8 ft) long, into a 3.7 m (12.1 ft) four-seater in seconds, thanks to a centrally located electric motor that "stretches" the car with the help of two mechanical screw-and-nut gears. The longitudinal members run on low-friction precision rollers, and disappear like a drawer into the rear of the floor pan.

According to Rinspeed, its engineers succeeded in designing this adjustable Presto floor pan with the necessary torsional rigidity of a roadster. To ensure that it is safe to use, the extension mechanism also features self-locking safety latches.

The contoured, drooping hood sports a mouth-like front grille whose upturned corners definitely appear to signal: "Hi, I feel good." The Presto's body tapers off into a front spoiler—a bumper in the traditional sense is omitted—and its headlamps come from the Mercedes C-Class coupe.

Instead of conventional lamp signals, the Rinspeed Presto informs other drivers in writing of the intention of its driver, using a powerfully bright LED lighting technology contributed by a specialist in advanced electronics. During braking, the international word "Stop" appears on the brake lights, and during turns "Turn" appears on the appropriate signal. With the lights turned on, the tail lights spell out "Presto" to anyone following.

The fiery-orange interior can be turned into a spacious pickup, or four-seat roadster, at the push of a button. The Presto betrays its magpie design heritage here too: the dash comes from the Mercedes-Benz A-Class, and the seats are from the Smart.

The car is not a likely candidate for production any time soon, but continues Swiss tuning company Rinspeed's tradition of providing auto-show concepts that get people talking … and get noticed, particularly in the tabloid newpapers.

Rover TCV

Here is MG Rover's attempt to prove that the British manufacturer has what it takes to compete in mainstream automotive markets. The TCV was created by design head Peter Stevens to demonstrate Rover's future-generation design direction. It claims to lead the beleaguered, and now independent, firm back to life after its enforced severance in 2000 from BMW.

The TCV mixes a classic station-wagon proportion with linear, geometrical styling features. At the front, the emphasis is placed on a "V" formation that emanates from the bumper, divides the grille and headlamps, and then continues to create a shoulder along the sides of the hood. In the metal, however, the frontal features vie for attention. Combined, they lack harmony, an accusation often leveled at Rover designs.

The side profile is sporty, and uncluttered, while the expanding roof rail at the rear, and the tailgate profile, share similarities to Patrick le Quément's Renault designs. The rear looks powerful, with a boldly shaped rear window, and large, trapeziodal tailpipes. The tailgate splits at the top of the bumper, which can then be folded down for easy access to the rear luggage space. The roof features longitudinal "Alpine light" windows, a common trend in concept cars today.

The TCV also looks as if it can perform off-road, with a little more ground clearance than a conventional wagon, appropriate wheels and tires, robust finishes on the lower body, and the potential for in-built traction-control technology.

This Rover "personality" could become a reality owing to MG Rover's wide-ranging new partnership with Far Eastern manufacturer China Brilliance, which is thought to include the development of the crucial mid-market models that the TCV previews.

ROVER

ROVER

Saab 9X

Design	Michael Mauer
Engine	3.0 V6 turbo
Power	224 kW (300 bhp) @ 5500 rpm
Torque	410 Nm (302 lb ft) @ 2200 rpm
Gearbox	6-speed sequential manual
Installation	Front-engine/all-wheel drive
Front suspension	Struts, lower A-arm, coil springs
Rear suspension	Independent, multi-link, coil springs
Brakes front/rear	Discs/discs
Front tires	245/40R19
Rear tires	245/40R19
Length	4156 mm (163.6 ins)
Width	1820 mm (71.7 ins)
Height	1365 mm (53.7 ins)
Wheelbase	2700 mm (106.3 ins)
Track front/rear	1570/1570 mm (61.8/61.8 ins)
Curb weight	1330 kg (2933 lb)
0–100 km/h (62 mph)	5.9 sec.
Top speed	250 km/h (155 mph)

Saab concept cars aren't just rare. With one other exception, there haven't been any. So when the wraps came off the 9X at the Frankfurt auto show in September 2001, the car world had to take note. When Saab, a company more usually credited with safety and environmental innovation, talks about the future of car design, it must be deadly serious.

The Saab 9X is a clever design, so versatile that it can be reconfigured as a high-performance coupe, a convertible, a wagon, or even a pickup. Saab genealogy, however, is immediately apparent. The dramatic, wrap-around windshield conceals the pillars, and is integrated with the door windows in an uninterrupted sweep of glazing. The 9X's high belt line, sloping roof, and bold grille are also unmistakably Saab.

The 9X is designed for all-weather driving, as well as for a variety of leisure, or load-carrying, needs. Its designers have ditched the traditional tailgate, hinged at the roof. A new, detachable, rear-roof rail is all that is required to attach the rear-roof panel, and locate the retractable rear-door window. Then the rear can be fully opened up. The design team has taken inspiration too from the sliding-floor feature of the Saab 9-5 Sportwagon, which extends the rear load space. With the roof structure removed, 9X would lose some of its inherent rigidity; for production, some redesign may be needed.

This is one of the most significant concept cars featured in this book. We are sure to see many more developments in adaptable design from other manufacturers.

What was that one other concept car? The ESV of 1985, a sporting two-seater with an all-glass roof. And there was, of course, the Saab 001, its first-ever prototype. In those days, building cars for "evaluation" meant something completely different: 001 took two weeks to build, and then covered 280,000 km (174,000 miles) of testing!

Saab 9-3X

Design	Michael Mauer
Engine	2.8 V6 turbo
Power	206 kW (280 bhp) @ 5500 rpm
Torque	400 Nm (295 lb ft) @ 1700–5500 rpm
Gearbox	5-speed semi-automatic
Installation	Front-engine/all-wheel drive
Front suspension	Struts, lower A-arm, coil springs,
Rear suspension	Independent, 4-link, coil springs
Brakes front/rear	Discs/discs
Front tires	245/40R20
Rear tires	245/40R20
Length	4380 mm (172.4 ins)
Width	1826 mm (71.9 ins)
Height	1498 mm (59 ins)
Wheelbase	2700 mm (106.3 ins)
Track front/rear	1580/1566 mm (62.2/61.7 ins)
Curb weight	1550 kg (3418 lb)
0–100 km/h (62 mph)	6.2 sec.
Top speed	250 km/h (155 mph)

An off-road Saab? Such a vehicle has yet to go into production, but one will come in the next few years in response to demand in North America, the Swedish company's biggest overseas market. As a guide to that car the 9-3X is only a hint, because the first production off-road Saab will be a "crossover" version of the 9-5 wagon.

A development of the Frankfurt 2001 concept, the 9X, the 9-3X is a further guide to the new language of Saab design under new styling boss Michael Mauer. Don't discount the possibility of a vehicle similar to the 9-3X appearing around 2005.

A recurrent theme in concepts for 2002 is "crossover." The 9-3X is exactly that—a coupe body profile on the high-ground-clearance body architecture of a 4×4 vehicle.

Distinctive Saab styling cues are present in the sweeping, wraparound windshield, and compact proportions, as well as tight wheel openings instead of the cavernous shapes usually employed on off-road designs. From the side, the proportion is rear-oriented, with the majority of the mass over the rear wheels. Together with a low hood, this gives a sleek appearance.

The wraparound screen is particularly important to the future of Saab design. A unique feature inspired by aircraft-canopy technology, it had to be dropped from Saab's recent production cars for cost-cutting reasons. Mauer's new design direction will revive that theme, in a classic case of a company rediscovering its design roots. The Saab 9-3X is the next step in the product plan, which will see at least one new car or concept announced every year for the next six years.

16.oct.01
26.oct.01

Seat Tango

Design	Steve Lewis
Engine	1.8 in-line 4 turbo
Power	132 kW (180 bhp) @ 5600 rpm
Torque	235 Nm (173 lb ft) @ 2100–5000 rpm
Gearbox	6-speed manual
Installation	Front-engine/front-wheel drive
Front suspension	MacPherson strut
Rear suspension	Longitudinal arms, with self steer
Brakes front/rear	Discs/discs
Length	3685 mm (145.1 ins)
Width	1714 mm (67.5 ins)
Height	1181 mm (46.5 ins)
Wheelbase	2200 mm (86.6 ins)
Track front/rear	1440/1429 mm (56.7/56.3 ins)
Curb weight	1150 kg (2536 lb)
0–100 km/h (62 mph)	7 sec.
Top speed	235 km/h (146 mph)
Fuel consumption	7.6 ltr/100 km (30.9 mpg)

The Seat Tango was conceived as an authentic spider, very much in the style of a Spanish Alfa Romeo. The resulting design of this attractive little car combines strength, simplicity, energy, and sportiness.

The Tango is expressive and innovative. The flowing lines of its compact body give a powerfully muscular impression. The strongly featured edge of the V-shaped hood sweeps upward to the door, and then continues rearward, falling slightly. The haunches are defined by powerful forms that run forward, stopping short at the doors. The V-shape is also visible at the rear, where the line falls downward to form the rear-lamp edge, the lid of the trunk, and, finally, the exhaust.

The passenger compartment is in tune with the exterior, with interlacing forms, materials, textures, and colors. The surface of the aluminum chassis is painted metallic gray that, in some areas, bleeds into the passenger compartment. Returning the compliment, the waterproof leather used to dress the dash, seats, backrests, and headrests overflows from the cockpit to cover the storage compartments located behind the seats.

Some essential parts of the structure have been deliberately left visible, such as the safety arch, the dash crossbeam, the undersides of the seats, and the steering column. They all have a rough, metallic finish, suggesting the aluminum bones of the car's skeleton. The controls, including the instrument-panel dials, pedal box, and footrests, are made from polished aluminum, a deliberately evocative throwback to 1950s and 1960s roadsters.

If Seat were to build the Tango, it would extend the desirability of the brand, which is now grouped by Volkswagen with its other sportily upscale brands such as Audi and Lamborghini. An exciting Mediterranean sports car would be created in the process.

TANGO

TANGO

TANGO

Skoda Superb

Design	Thomas Ingenlath
Engine	2.8 V6 (2.0, 1.8 turbo, 1.9 diesel, and 2.5 turbo-diesel, all in-line 4, also offered)
Power	142 kW (190 bhp) @ 6000 rpm
Torque	280 Nm (207 lb ft) @ 3200 rpm
Gearbox	5-speed automatic Tiptronic/5-speed manual/6-speed manual
Installation	Front-engine/front-wheel drive
Front suspension	Multi-link independent
Rear suspension	Coil springs and dampers
Brakes front/rear	Discs/discs, ABS, ESP
Front tires	225/45R17
Rear tires	225/45R17
Length	4803 mm (189.1 ins)
Width	1765 mm (69.5 ins)
Height	1469 mm (57.8 ins)
Wheelbase	2803 mm (110.4 ins)
Track front/rear	1515/1515 mm (59.6/59.6 ins)
Curb weight	1501 kg (3310 lb)
0–100 km/h (62 mph)	8 sec.
Top speed	237 km/h (147 mph)
Fuel consumption	9.9 ltr/100 km (23.7 mpg)
CO_2 emissions	238 g/km

Peugeot's 406, Renault's Laguna, and other established upper-medium-sector cars have an unexpected new rival from Skoda. The Superb is the Czech brand's flagship, a class of car not offered by Skoda since the 1930s. The last bearer of the name was the 1938 Superb, a spacious limousine with a lavish interior that turned out to be short-lived, since World War II began shortly afterwards, followed by the Soviet era and Skoda was then obliged to serve strictly proletarian fare for some fifty years.

The new model has a classic, three-compartment body that shares its overall shape with the Volkswagen Passat. The relationship goes deeper than that, because the Superb also shares the Passat's excellent drivetrain platform. However, it is differentiated by such key features as the characteristically rectangular radiator grille, large, clear-optic headlamps, chrome-edged side windows, and the chrome strip that forms a ring around the car at bumper height. It is sober and conventional rather than mold-breaking.

The interior uses the high-quality materials found in modern Volkswagens. It is an uncommonly roomy sedan—Skoda says the most spacious in its class—particularly in the back, even for tall passengers, thanks to the Passat-inspired long wheelbase.

The Superb is another giant step forward for Skoda. The brand has undergone an unparalleled transformation in the last ten years, thanks to Volkswagen's corporate patience and product-development fastidiousness. By using Volkswagen parts, Skoda saves costs while producing quality cars. The biggest challenge now for Volkswagen is to maintain design differentiation. Otherwise, excellent Skodas such as the Superb could easily take sales away from its own core brand.

MBN 70-55

V6 30V

Skoda Tudor

Design	Thomas Ingenlath
Engine	2.8 V6
Power	142 kW (190 bhp)
Installation	Front-engine/front-wheel drive
Wheelbase	2803 mm (110.4 ins)

Skoda's Tudor study—the name, an Americanization of "two-door," has nothing to do with British royalty—is a break from tradition. This elegant coupe is a car you would normally expect to see with Alfa Romeo or Mercedes-Benz emblems. Skoda says it is intended to show the depth of creativity of Skoda designers, and to lift the curtain on behind-the-scenes car development.

Derived from the new Skoda Superb (see p. 228), this four-seat coupe combines elegance with sportiness and excellent performance. It adheres to key Skoda values: solidity and prestige.

The wedge-shaped contour of the body, emphasized by two smoothly ascending lines, is apparent when viewed from the side. The second lower line, running from the door handles toward the rear wheel arches, ends above the rear light. It creates a sporty character with blistered wheel arches.

The interior uses onyx and ivory colors, like those in the Superb. A horizontal strip of matt-finish aluminum decorates the panel, together with aluminum instrument details designed as traditional chronographs—more sportiness.

The Tudor is a fully functional car using only contemporary, realistic technology. Skoda says it is not considering it for production, and that it is experimental, but it does seem a suitable candidate for the Czech production lines. Expect more exciting products from Skoda soon, as the brand grows ever stronger under Volkswagen's leadership.

Subaru Baja

Design	Peter Tenn
Engine	2.5 in-line 4
Power	123 kW (165 bhp) @ 5600 rpm
Torque	225 Nm (166 lb ft) @ 4000 rpm
Gearbox	5-speed manual/4-speed automatic
Installation	Front-engine/4-wheel drive
Front suspension	Strut type with lower L-arms, liquid-filled L-arm rear bushings
Rear suspension	Multi-link, radius arms
Brakes front/rear	Discs/discs, ABS
Front tires	225/60R16
Rear tires	225/60R16
Length	4910 mm (193.3 ins)
Wheelbase	2649 mm (104.3 ins)
Curb weight	1588 kg (3502 lb)

Inspired by surf-bum culture on the west coast of California, the Subaru Baja is the production version of the ST-X concept shown at the Los Angeles Auto Show in 2000. The bizarre genesis of the Baja was the propensity of impecunious surfers to drive beat-up sedans which, unconcerned by weather or security, they then modified by smashing out the rear windshield to carry their surfboards. Why not incorporate a similar idea in a volume production model? reasoned Subaru.

The Baja is based on the Subaru Outback platform. Instead of a covered station-wagon load bay, it has an open, pickup-truck bed. Where the Baja scores is in its extra flexibility, thanks to Subaru's new Switchback seating system. The backs of the rear seats also constitute the rear wall of the cabin. When the seats are folded away, the rear cabin becomes part of the load bay, making room to carry a surfboard or two without having to resort to smashing the rear window.

This is the second vehicle to come to market with a similar idea—Chevrolet's Avalanche is the other. The Baja is a smaller vehicle than the large Avalanche. Because it is based on a unitary-construction car platform, it will be sportier to drive and more refined.

Nevertheless, it will be marketed with a strong accent on the outdoors. The Baja's expandable bed can accommodate mountain bikes, ski or snowboard gear, and scuba kit. Useful features such as sport bars, a bed light, and four tie-down hooks are incorporated into the bay. Roof rails, and crossbars, are part of the design, and attachments for skis, kayaks, and snowboards are available.

SUBARU
FJR 231

Suzuki Aerio

Engine	2.0 in-line 4
Power	105 kW (141 bhp) @ 5700 rpm
Torque	183 Nm (135 lb ft) @ 3000 rpm
Gearbox	5-speed manual/4-speed automatic
Installation	Front-engine/front-wheel drive
Front suspension	MacPherson strut
Rear suspension	MacPherson strut
Brakes front/rear	Discs/drums
Front tires	195/55R15
Rear tires	195/55R15
Length	4229 mm (166.5 ins)
Width	1720 mm (67.7 ins)
Height	1549 mm (61 ins)
Wheelbase	2479 mm (97.6 ins)
Track front/rear	1450/1445 mm (57.1/56.9 ins)
Curb weight	1210 kg (2668 lb)
Fuel consumption	8.8 ltr/100 km (26.6 mpg)

Small cars do not sell as well as big 4×4s in North America, but they do help with two major corporate targets. One is to help balance a car company's average fleet fuel economy, called CAFE, a US Federal Government requirement. For Suzuki that is important, since most of its US sales are of thirstier 4×4s. The second, and equally important, target is to attract younger buyers, who will become customers for more expensive models in the future.

These two criteria have made the Ford Focus and Toyota Echo (Yaris in Europe) big hits in the US. Not surprisingly, Suzuki is following suit with its Aerio, a small sedan, whose styling was previewed at the 2001 New York auto show.

Like its rivals, the Aerio has a distinctive shape, aimed at appealing to buyers wanting to make a contemporary style statement. Punctuating the Aerio's exterior are bold front and rear bumpers, sporty side moldings, wheel flares, and a striking front grille, giving both models an urban, edgy feel. The distinctive theme continues with the multi-reflector, jewel-type headlamps, and the prism tail lamps.

"Walk-through" instrument panels—those that don't occupy space between the driver and the passenger—are increasingly a feature in interior design. These are made possible by the upright body architecture of many of today's small cars. The Aerio follows this trend. In appealing to youthful buyers, it also adopts a digital speedometer and tachometer—features similar to the Toyota Echo/Yaris. The Aerio is more conservative with its panel design, however, positioning the instruments in a conventional spot in front of the driver, rather than on a central binnacle as in the Echo/Yaris.

Suzuki SX

Engine	2.0 in-line 4 supercharged
Power	164 kW (220 bhp)
Gearbox	5-speed manual
Installation	Front-engine/4-wheel drive
Front suspension	MacPherson strut
Rear suspension	MacPherson strut
Brakes front/rear	Discs/discs, ABS
Front tires	225/45R17
Rear tires	225/45R17
Length	4201mm (165.4 ins)
Width	1880 mm (74 ins)
Height	1580 mm (62.2 ins)
Wheelbase	2479 mm (97.6 ins)
Track front/rear	1631/1659 mm (64.2/65.3 ins)

As a "Sports Crossover" car, to use Suzuki's own terminology, the SX is something of a disappointment. However, rather than starting from scratch, Suzuki's design team has based the car on the Liana, a five-door hatchback already in production and on sale. On that premise, Suzuki has done wonders in transforming a drab family car into a dashing little runabout.

Where the Liana looks staid at the front, the SX exudes sportiness, even if the end result appears somewhat inconsistent. The aluminum "mouth" on the front bumper, for instance, together with the sill and wheel-arch extensions, look like parts of an after-market body kit, rather than being purpose-designed for the SX.

At the rear, Suzuki has similarly tortured the surfaces, rather than simplifying them, and thus making the car's design more understandable. The sporty aspirations are defined by large bumper and sill moldings, blistered wheel arches, and a tailgate-mounted spoiler. The SX features contrasting panels of metallic paint on its bumper inserts and door handles, distinct prism-type headlamps, and vertical fog lamps to give it a high-tech look.

The pearlescent body paint is designed to turn a reddish hue while the car is in the shade, and a strikingly bright yellow while out in direct daylight, a changeable color scheme inspired by Suzuki's 1970s motocross cycles.

Inside, sporty black-and-yellow leather seats, and yellow bands that highlight the door trims, dash panel, and instrument binnacle continue the sensory assault. This bright and contrasting interior creates a youthful, sporty image for the SX that no Liana has ever possessed.

Suzuki claims that the SX draws heavily on the company's sporting heritage, particularly with the supercharged engine under that vividly styled hood. However, a car such as this proves the difficulty of using design and imagery to link practical family cars with stripped-down racing motorcycles.

SUZUKI
SUZUKI

Th!nk City

Design	Stig Olav and Katinka von der Lippe
Engine	Three-phase AC induction motors, 110 or 220 V NiCd battery
Power storage	11.5 kWh, 100 Ah
Gearbox	Single-speed, gear-reduction drive
Installation	Front-wheel drive
Brakes front/rear	Discs/drums
Front tires	155/70R13
Rear tires	155/70R13
Length	2991 mm (117.8 ins)
Width	1604 mm (63.1 ins)
Height	1560 mm (61.4 ins)
Curb weight	941 kg (2075 lb)
Top speed	90 km/h (56 mph)
0–49 km/h (30 mph)	7 sec.
Range	53 miles (85 km), urban driving
Full charge time	6–8 hours

When former Ford CEO Jac Nasser embarked on his turn-of-the-millennium buying spree—to support his vision of Ford as a diversified company providing all kinds of services connected with transport and mobility—one of his most dramatic purchases was the Norwegian company Th!nk. Th!nk was small, and in financial difficulties, but was well respected for its pioneering City, an electric, urban two-seater.

Nasser's vision was to integrate Th!nk into his empire as Ford's pro-environmental brand. The small City model would be just the beginning, he reasoned; larger cars would follow, and existing ranges could even include low- or zero-emission Th!nk derivatives.

It was clear from the start, however, that the original City, with its matt, plastic body moldings, and toy-like looks, would have to be updated to be commercially competitive. This new version, which carries a prominent Ford emblem on the tailgate, is the result of that rethink. It comes across as much smarter, and more grown up, although some panels still lack the gloss found on similar cars, such as the Smart. Nevertheless, the visible aluminum upper spaceframe, which forms the windshield pillars and the cantrail above the windows, is a more stylish element.

Inside, the City is improved, but still falls some way short of the comfortable, and fashionable, aura provided by the Smart. There is adequate space for two people, and, considering the overall dimensions, the trunk is reasonable. The rear view, with the dark, smoked-glass tailgate extending right down to bumper level, is perhaps the least successful aspect of the redesign.

There are new are features aimed at the American market, such as power steering, power windows, and power locks, dual cup-holders, air conditioning, and driver and passenger air bags.

The AC electric motor drives the front wheels, and gives a top speed of 90 km/h (56 mph), and a range of over fifty miles on a full battery. Five Ford dealerships in California are set to lease the cars to fleet and retail customers within thirty-five miles of their location, and a leasing deal for private motorists has just been announced for London.

TH!NK
TH!NK

Toyota ccX

Engine	2.4 in-line 4
Gearbox	4-speed automatic
Installation	Front-engine/front-wheel drive
Brakes front/rear	Discs/discs
Front tires	225/45R18
Rear tires	225/45R18
Length	4331 mm (170.5 ins)
Width	1791 mm (70.5 ins)
Height	1445 mm (56.9 ins)
Wheelbase	2601 mm (102.4 ins)
Track front/rear	1542/1524 mm (60.7/60 ins)
Curb weight	1275 kg (2811 lb)

Toyota sells nearly one in ten of all the cars bought in the United States, but it can't afford to miss any emerging market niches. Hence the ccX (Concept Coupé Crossover), a vehicle in the mould of the Acura RD-X and Saab 9-3X (see pp. 20, 224).

The shape has an aggressive, sports-coupe look, defined by strongly arched character lines, sharp surface edges, a tall, sloping roof line, and rounded front and rear contours. Meanwhile, the roof comprises two large, powered sunroofs, each made up of four glass panels, similar to those fitted on the Audi A2. These panels tilt individually, and slide forward or backward, creating large openings over the passenger compartment and cargo area. An equally novel use of glass—echoes of the Nissan Quest (see p. 188)—for the tailgate allows each of the two doors to retract into the bumper and out of the way, giving access to the rear cargo area. Convenience is addressed with a hand-sensitive system that automatically opens the doors.

In the cabin, the dominant theme is the instrument cluster, centrally mounted on the dash—a design theme already in production on the Yaris supermini. The blue instrument pack is part of a color scheme carried through to the border of a 178 mm (7 ins) multi-display monitor for DVD entertainment, navigation, and audio systems. A remote-control panel for the monitor, and the cellphone, is located on the center console, protected by a blue, acrylic, sliding, cover.

The ccX is an aggressive-looking coupe (similar in many ways to the Voltz [see p. 258]) that combines the styling of a sports coupe with some cargo-carrying capability.

Toyota Corolla

Design	Takeshi Yoshida
Engine	1.8 in-line 4 turbo (1.4, 1.6, and 1.8 in-line 4, and 2.0 in-line 4 diesel, also offered)
Power	141 kW (189 bhp) @ 7800 rpm
Torque	180 Nm (133 lb ft) @ 6800 rpm
Gearbox	5-speed manual/4-speed automatic
Installation	Front-engine/front- and 4-wheel drive
Front suspension	MacPherson strut
Rear suspension	Torsion beam
Brakes front/rear	Discs/discs
Front tires	195/55VR16
Rear tires	195/55VR16
Length	4180 mm (164.6 ins)
Width	1710 mm (67.3 ins)
Height	1475 mm (58.1 ins)
Wheelbase	2600 mm (102.4 ins)
Track front/rear	1480/1460 mm (58.3/57.5 ins)
Curb weight	1255 kg (2767 lb)
0–100 km/h (62 mph)	8.4 sec.
Top speed	225 km/h (140 mph)
Fuel consumption	7.6 ltr/100 km (37.1 mpg)
CO_2 emissions	181 g/km

Toyota's new Corolla is now in its ninth generation, with a new incarnation of the brand appearing around every four years. The latest one is part of Toyota's well-defined European product strategy, which has recently brought us the Yaris, and the MR2.

The Corolla has a clear family connection with the smaller Yaris, thanks to its European design signature, and key features such as the new headlamps, fitted with clear lenses to give a jewel-like effect, while the small bulge at the front of the hood draws the eye toward the Toyota emblem on the grille.

The overall design is fairly conventional, and aims for mass appeal. The three- and five-door hatchbacks are the core models, and feature sporty styling, particularly the three-door with its large bumpers, roof-mounted spoiler, and dynamically shaped rear-quarter glass.

The new product range includes the Verso, which is a compact MPV that shares its design cues with the other Corolla models, thus developing a strong family identity. In addition, Toyota also offers an estate-car version as a practical, stylish solution for customers who appreciate the traditional nature of a wagon. The Corolla sedan offers a reasonably elegant choice for more conservative customers.

Inside, the instrument panel and trim have a soft-touch quality—Toyota claims the operation of switches and dials is carefully tuned: for example, the weighting and closing sounds of locks and doors have been calibrated to give the Corolla a "premium" feel.

Toyota's enviable success is built upon a long history of building good-quality, reliable cars. So the Corolla, while, perhaps consciously, never the most cutting-ege of designs, has an immensely loyal following worldwide, that will go some way to counter the threat of considerable rivalry.

COROLLA
COROLLA

Toyota DMT

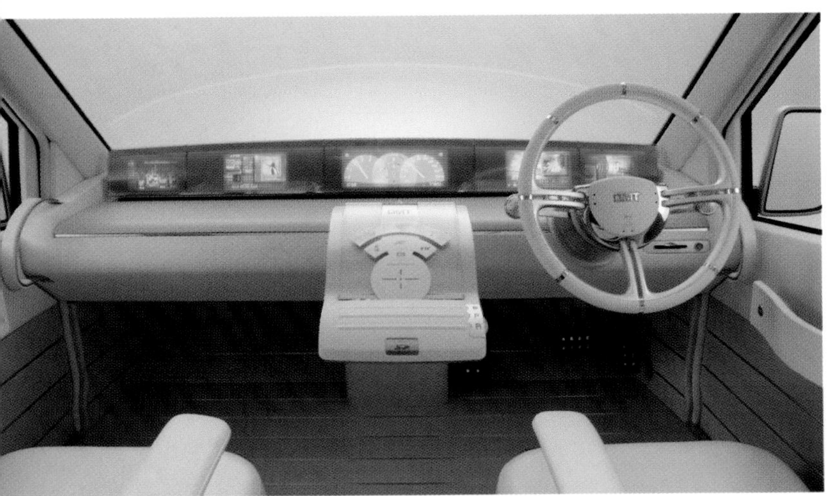

Engine	2.4 in-line 4
Gearbox	Automatic
Installation	Front-engine/front-wheel drive
Front suspension	MacPherson strut
Rear suspension	Torsion beam
Brakes front/rear	Discs/discs
Length	4850 mm (190.9 ins)
Width	1825 mm (71.9 ins)
Height	2030 mm (79.9 ins)
Wheelbase	2900 mm (114.2 ins)
Track front/rear	1600/1600 mm (63/63 ins)

There is a disjointed look to this hefty-looking MPV, which Toyota has christened the Dual Mode Traveller (DMT). That is entirely intentional. The DMT's silhouette features a high-lift cabin, that emphasizes the driving space, while the window surrounds differentiate the bizarrely named front "drive-mode" and rear "stay-mode" areas. The belt line, deliberately not flowing from front to rear, visibly separates the two interior segments.

The protruding cab is the "drive-mode" portion, offering a high seating position for two, with commanding forward visibility. A large, slatted roof, which, when open, provides an airy, relaxing environment, sits above the raised cabin. The interior is simple, and spacious; digital instruments sit in a wide monitor in front of the driver.

The rear, the "stay-mode" area, is an uncluttered box with a low, flat floor, an interior that could be adapted easily into an office, a studio, a workshop, even a laboratory, if necessary. It is partitioned from the "drive-mode" area, but is easily accessed by a linking door and steps.

Admittedly, it is essentially a large van, but the DMT splits the dual functions of on-road traveling and on-site versatility, and you quickly begin to imagine a thousand uses for it. After all, this is an area that neither large MPVs nor converted delivery vans have addressed so concisely, although the small worldwide market for such a specific, and expensive-to-engineer, vehicle hints at the DMT's fate as a mildly interesting sideshow.

DMT

DMT

DMT

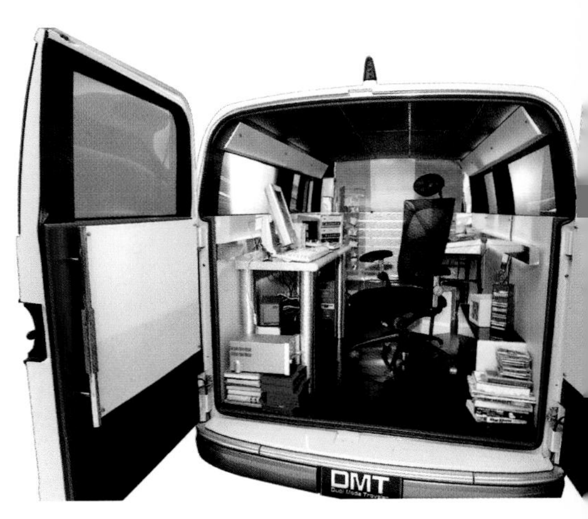
DMT

Toyota ES³

Design	Kazumiko Miyadera
Engine	1.4 in-line 4 turbo-diesel
Power	74 kW (100 bhp)
Torque	170 Nm (125 lb ft) @ 2000 rpm
Gearbox	Automatic CVT
Installation	Front-engine/front-wheel drive
Front suspension	MacPherson strut
Rear suspension	Trailing arm, cross beam
Brakes front/rear	Discs/drums
Front tires	145/70R15
Rear tires	145/70R15
Length	3520 mm (138.6 ins)
Width	1630 mm (64.2 ins)
Height	1460 mm (57.5 ins)
Wheelbase	2310 mm (90.9 ins)
Track front/rear	1450/1440 mm (57.1/56.7 ins)
Curb weight	700 kg (1543 lb)
Fuel consumption	2.7 ltr/100 km (86.9 mpg)

The quest for ultimate aerodynamics on road cars has gone quiet in recent years, so the development of Toyota's Eco Spirit Cubic, or ES³, is welcome news. Pininfarina's Studio CNR of 1977 showed that a drag coefficient (Cd) of just 0.201 could be achieved for a roomy family sedan. Five years later, the substantial Audi 100 went on sale with a Cd of just 0.30.

Toyota has chosen to create a compact, sporty, yet environmentally friendly vehicle in the ES³, and has managed an exceptionally aerodynamic Cd of just 0.23 for its exterior. It has achieved this in a conventional, Yaris-sized, three-door hatchback with a rounded nose, small cooling intakes, non-protruding headlamps, and smooth wheel covers. The roof line drops, and the body waists inwards at the rear, to reduce turbulence.

Toyota says ES³ consumes only 2.7 liters of fuel every 100 km (104.7 mpg), thanks to its wind-cheating details and super-low 700 kg (1544 lb) weight, achieved by using a combination of aluminum and plastic for its body. Toyota insisted on a high degree of recyclability. The plastic panels used in the back-door windows, the rear floor, the radiator supports, the side panels, the bumpers, and the fuel tank can all be reprocessed.

There are energy-conserving tricks under the skin, too. ES³'s 1.4 ltr diesel engine is mated to a continuously variable transmission. A "stop and go" management system turns the engine off when the car is stationary. A braking-energy regenerating system, one of Toyota's key hybrid technologies, is used to convert vehicle-deceleration energy into electrical energy. This is then stored in a capacitor, and used for auxiliary electrical tasks, such as restarting the engine.

As an advanced concept prototype, the Toyota ES³ could indeed be more "slippery." For such a small car, however, it couldn't be much more thrifty, or practical.

Toyota FXS

Engine	2.4 in-line 4
Gearbox	Automatic
Installation	Front-engine/front-wheel drive
Front suspension	MacPherson strut
Rear suspension	Torsion beam
Brakes front/rear	Discs/discs
Length	4153 mm (163.5 ins)
Width	1872 mm (73.7 ins)
Height	1110 mm (43.7 ins)
Wheelbase	2502 mm (98.5 ins)
Track front/rear	1549/1560 mm (61/61.4 ins)

Toyota's Future Experimental Sports car concept, the FXS, was designed to have a simple and sexy form with presence. The FXS is fitted with a 4.3 ltr engine, a six-speed transmission, and double-wishbone suspension that promises driving and handling excitement. Its launch coincides with Toyota's entry into the 2002 Formula 1 world championship.

The exterior is composed of emotional and retro surfaces, with high wheel arches, and a drooping belt line that emphasizes the openness of the cockpit. The small windshield and door windows demonstrate Toyota's intention that the FXS be seen as a classic sports car, with more than a hint of the Toyota 2000GT of the 1960s—an open version of which starred in the James Bond film *You Only Live Twice*. This timeless look is enhanced by the low-slung body with its barreled doors, and the choice of chrome for the wheels.

Toyota says the design stresses three points: a 50:50 balance between front and rear—helped by seating the driver a long way back in a low, and stretched-out, position—good directional agility, and a low center of gravity. These are important for creating the high performance and stability a sports car needs when driven to its limits. The interior of the FXS is driver-focussed, furnished in full leather, with clever chrome highlights. It has simple instruments, with blue-white illuminated dials set in a deep-blue background.

The FXS is clearly a car for those who like driving. Toyota won't comment on whether, and when, the FXS might be built, but if the company's entry into the world of Formula 1 is a success, the market will be ripe for Toyotas that are more obviously sporting cars than the current Celica and MR2 models.

FXS

Toyota Ist

Engine	1.3 in-line 4
Installation	Front-engine/front-wheel drive
Length	3855 mm (151.8 ins)
Width	1695 mm (66.7 ins)
Height	1530 mm (60.2 ins)
Wheelbase	2370 mm (93.3 ins)
Track front/rear	1450/1450 mm (57.1/57.1 ins)
Curb weight	1435 kg (3164 lb)

The key elements Toyota used to define the design of the Ist were fitness, intelligence, and elegance. It has largely succeeded with all three, coming up with an attractive, and sporty, potential rival to the Volkswagen Golf, should the Ist get the go-ahead for production.

Sporty wheels positioned at each corner, wrapped by body-colored surfaces, and large, jewel-like headlamps, emphasize the Ist's sense of purpose. Subtle feature lines, and the dark-tinted windows that begin at the windshield and continue right around the car, give it an "intellectual" quality, while its continuously flowing lines make it look quite elegant. A short, steeply raked hood generates extra volume at the front, and blends gently with the windshield. This creates more space inside the car, and better aerodynamics. But the blistered wheel arches look slightly clumsy, and out of proportion with the overall size of the car.

The Ist's interior uses simple geometry, based on parallels, right angles, and curves that extend from front to rear. Geometric patterns, and metallic details, are used to create a more sophisticated look.

Toyota has a strong reputation for excellent-quality, owner-friendly cars. The dashing Ist concept adds some sporting excitement to everyday practical demands, and looks feasible as a production model in this mainstream sector. It would sit comfortably within Toyota's existing product range, somewhere between the top of the Corolla range and the bottom of the European Avensis series. If the Ist is as good to drive as it looks, Toyota would be wise to produce it.

7TH AVE.
PARK AVE.
7TH AVE.
NO PARKING IN THIS BLOCK
NO COMMERCIAL TRAFFIC
NO LEFT TURN
CAUTION
PEDESTRIANS
ist

Toyota pod

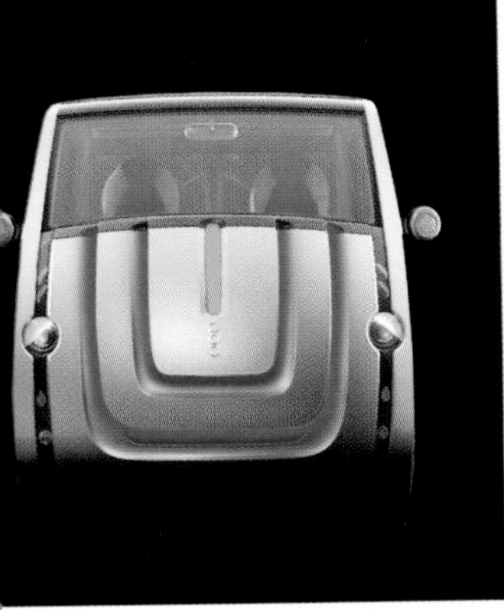

sad

angry

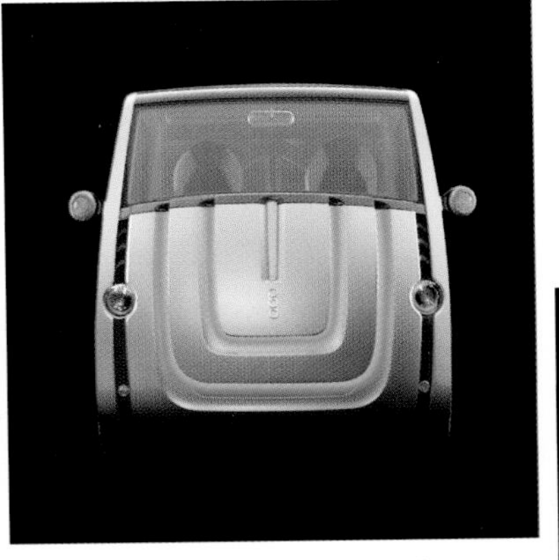

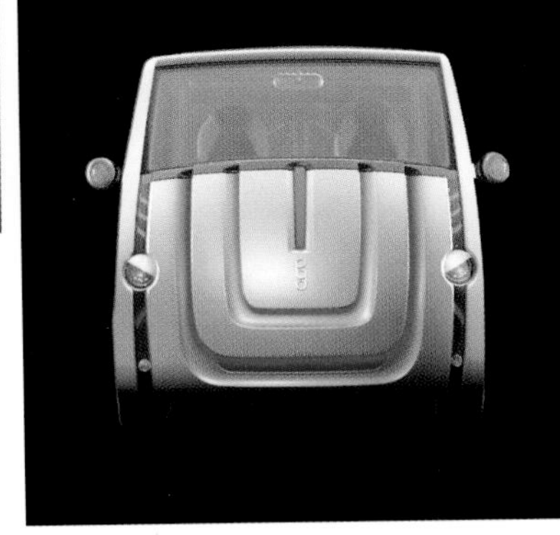

happy

sleepy

"What your car says about you"—this idea has been worn smooth by the marketing people over the years. Yet the Toyota pod, one of the most talked-about concepts unveiled at recent auto shows, does just that. There are some similarities with Toyota's offbeat WiLL Vi production model, but the pod is unique in that it can "express" the driver's—and its own—feelings.

The pod, jointly developed by Toyota and Sony, begins to explore the potential for communication between people and their cars. Toyota says that it "is devised as a partner, sharing your moods, and growing with you, like your family and friends. Technological innovations enable the pod to show emotional states, and learn from experience."

The pod adopts a cheerful expression when its owner approaches, an angry demeanor when the driver brakes hard, or swerves sharply, and a sad expression if a tire punctures, or it runs out of fuel. The tail-like antenna at the rear can also wag to express feelings.

The front of the pod can express ten different emotional states, including happiness, surprise, and sleepiness. It does this by using different LED colors, and a combination of its front features: its eyes (the headlamps), its mouth (the grille), and its ears (the door mirrors), all of which light up as appropriate. Its attractively curvy exterior is identical at front and rear.

Inside the pod, each seat can be swiveled around to create a sociable, inward-facing area. The car is also fitted with monitors so that passengers can enjoy music or videos individually.

Emotions aside, the pod's highly decorated exterior, when fully lit, could be very confusing to fellow road users. Then again, the pod's design is purely conceptual. If Toyota and Sony continue their partnership and find a way to bring a car like this to market, it could turn our attitude to our automobiles on its head.

Design	Simon Humphries
Engine	1.5 in-line 4
Installation	Front-engine/front-wheel drive
Length	3930 mm (154.7 ins)
Width	1800 mm (70.9 ins)
Height	1860 mm (73.2 ins)
Wheelbase	2500 mm (98.4 ins)
Track front/rear	1460/1440 mm (57.5/56.7 ins)

Toyota RSC

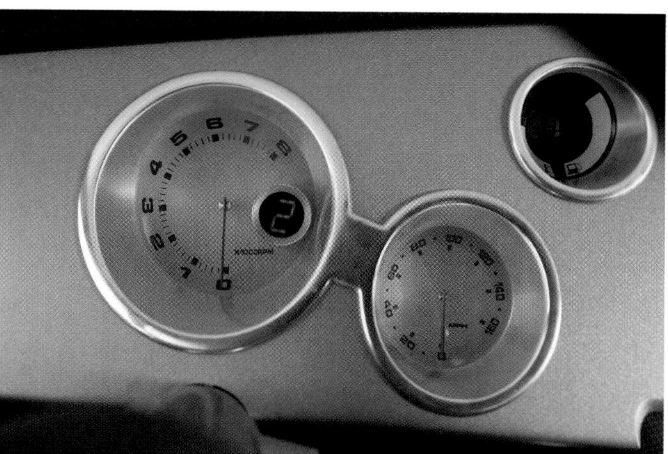

Design	Kevin Hunter
Gearbox	6-speed, sequential manual
Installation	4-wheel drive
Length	4115 mm (162 ins)
Width	1850 mm (72.8 ins)
Height	1550 mm (61 ins)
Wheelbase	2490 mm (98 ins)

Rugged Sports Coupé (RSC) is purely a visual concept, developed by Toyota as a could-be sports car with lots of appeal to young rallying fanatics. It will almost certainly never see a factory floor; such a car would occupy a specialist niche, with few actual buyers, even though its admirers would see it as the ultimate in automotive hedonism. A few of its design brush strokes, though, may just be fed through to production cars.

Inspired by the sort of wild machines that battle it out in the World Rally Championship, and the grueling Paris–Dakar rally-raid, the RSC has an aggressive exterior made up of angular surfaces and swaggering performance bulges. It gives the impression of a car that has been bolted together by a race team, rather than assembled in a factory. The chunky wheels, front sump guard, and high ground clearance suggest that the RSC will be more than happy plugging mud in the most difficult of competition terrain.

The interior is simple and practical, with lots of satisfying race-car touches. The liberal use of aluminum suggests that this car has been designed for light-weight agility and handling, with interior comfort a distant third in the priority list. The instrument panel also gives the impression that the RSC was built in a race shop, the large metal faceplates on the round meters symbolizing accuracy, and robustness. The sequential gear stick is high-mounted, just as in a rally car. The image is completed by a global positioning system, and lightweight, carbon-fiber racing seats fitted with full body harnesses.

TOYOTA

Toyota UUV

Design	Toyota European Design Center, France
Engine	2.0 in-line 4
Installation	Front-engine/4-wheel drive
Length	4430 mm (174.4 ins)
Width	1820 mm (71.7 ins)
Height	1650 mm (65 ins)

The Urban Utility Vehicle (UUV) concept highlights Toyota's spirit of innovation, and constant desire to design exciting new models. It mixes 4×4 elements, such as strongly flared wheel arches and large wheels, with typical "urban" features. It also seeks to blend Japanese culture—technical innovation—with European culture—the exterior shape.

The exterior features strong, sweeping character lines, sharp surface edges, and a tall, sloping roof line. Tense geometric surfaces typify the urban environment in which this car will mainly be used, yet the overall effect is sleek and aerodynamic.

The UUV is in the vanguard of a new generation of "crossover" cars that blend the best SUV characteristics with premium-car values such as design, road behavior, comfort, and performance. The body style, for instance, is a hybrid hatchback/sedan/wagon.

Who will buy one? Target UUV customers are part of a progressive, successful demographic group who already choose large family cars for their comfort, power, equipment, and prestige.

The interior features a full-width "Glass Vision" screen that offers entertainment and navigation functions to driver and passenger. The screen is made of a photopolymer, for a more realistic display and large projection area.

Is Toyota's UUV really a glimpse of the future?

01:25:4

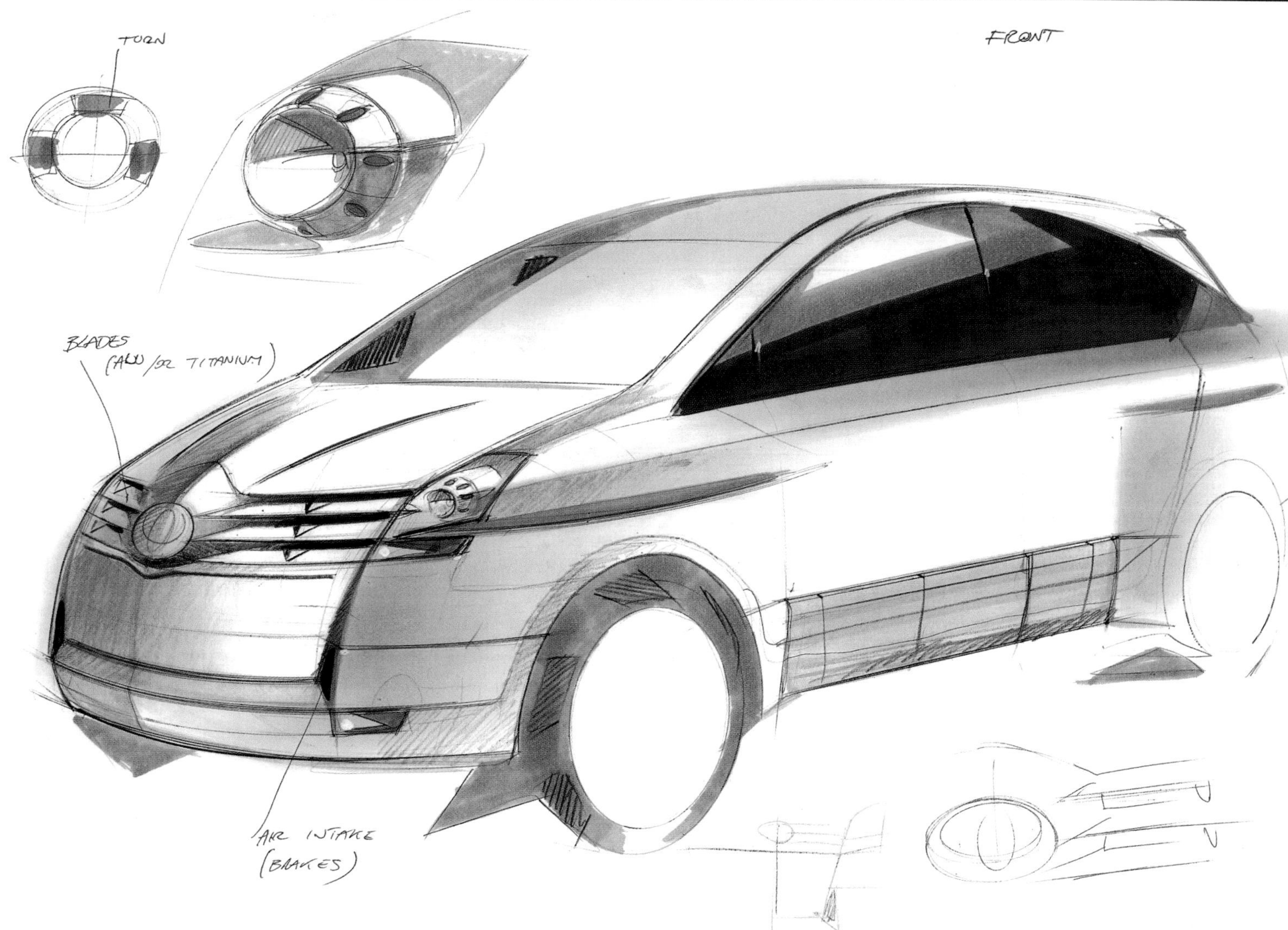
TURN
FRONT
BLADES
(ALU/OR TITANIUM)
AIR INTAKE
(BRAKES)

Toyota Voltz

Design	Kevin Hunter
Engine	1.8 in-line 4
Gearbox	6-speed manual
Installation	Front-engine/front-wheel drive
Length	4365 mm (171.8 ins)
Width	1775 mm (69.9 ins)
Height	1600 mm (63 ins)
Wheelbase	2600 mm (102.4 ins)
Track front/rear	1510/1520 mm (59.4/59.8 ins)
Curb weight	1300 kg (2866 lb)

The slightly sinister-looking Voltz is a sports-utility wagon designed for people with a youthful lifestyle. Toyota claims that this "crossover" car has the space of a minivan, the sports features of a coupe, and the power, and practicality, of an SUV—a tricky mixture to get right. It is wrapped up in an awkward-looking, but attention-grabbing, body, whose surfaces and features often fail to flow with each other.

The Voltz looks aggressive at the front, with a large, pointed, dark-colored grille, and headlamps that cut into the bumper, and hood. The coupe-style sloping roof line is emphasized by the rising belt line on the rear door. The roof slope seems to compromise the available space inside the Voltz, creating an impractically small tailgate opening. It is a shape Honda has achieved more usefully in its new Civic, although that production car does not have such grandiose aspirations as the Voltz.

The cabin color is mainly black, with a metallic center cluster, chrome trim, and four Optitron dials that project a sporty, and technical, image. The rear parcel shelf, made from tough plastic, can pivot rearward and downward, providing a working surface above the rear bumper. With rear seats folded, the Voltz can accommodate a mountain bike—with the front wheel removed—slid in on a frame designed rigidly to support it.

Despite its name, the Voltz uses a conventional gasoline engine and not an electric motor.

VOLTZ

VOLTZ

Toyota WiLL VC

Engine	1.3 in-line 4
Installation	Front-engine/front-wheel drive
Length	3695 mm (145.5 ins)
Width	1675 mm (65.9 ins)
Height	1535 mm (60.4 ins)
Wheelbase	2370 mm (93.3 ins)
Track front/rear	1440/1420 mm (56.7/55.9 ins)

In 1999, Toyota decided to create an all-new brand to appeal to those young consumers who've been unimpressed by the product "establishment" that Toyota exemplifies in Japan. But it had in mind more than a car brand: the WiLL project is a collaboration with several non-automotive companies, including Panasonic and Ashii Beer. The aim was to create a new brand that could offer a wide range of products, all marketed under the WiLL name, each throwing away the rulebook for its sector.

The WiLL Vi four-door car has already been on sale for a year or two, and is reportedly a winner. Thus the cheeky VC aims to build on that left-field achievement.

The styling of its bubble-shaped exterior revolves around a "cyber capsule" theme. It uses many different curves to define it, the main ones being the rounded hood, and large wheel arches, which also give it a sense of fun. Its execution is rather incoherent, jumbling sporty elements such as a rising belt line, and six-spoke wheels with inconsistent proportions, intersecting forms, and a gawky mix of feature lines.

Inside, the VC's design repeats the circular theme, exaggerating the car's unique identity. A so-called "G-Book" terminal customizes data that the driver feeds into it. It can also be used to access the Internet, receive or send emails, and get real-time navigation information. Its two-way support allows voice recognition, the reading aloud of received information, and hands-free phone calls.

Toyota says the WiLL VC represents a futuristic commuter vehicle, but whether Toyota and its WiLL partners will add it to their eclectic catalog remains to be seen. WiLL shouldn't be taken lightly, however: in 1988 Toyota created another brand from scratch—Lexus—and just look at that today.

WiLL

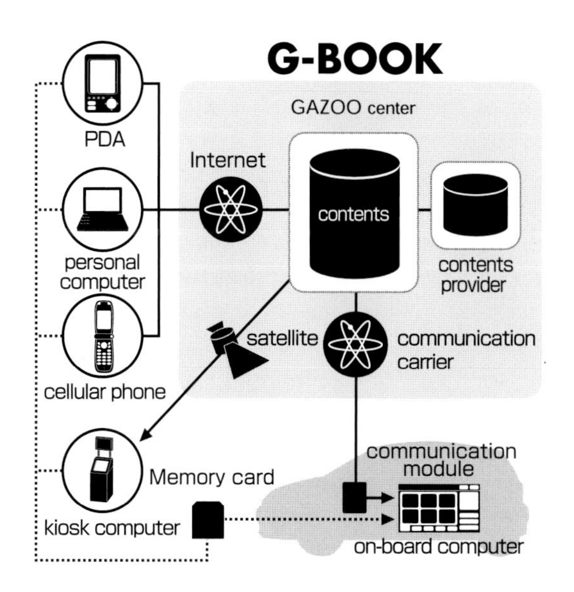
G-BOOK
GAZOO center
PDA
Internet
contents
contents provider
personal computer
satellite
communication carrier
cellular phone
communication module
Memory card
kiosk computer
on-board computer

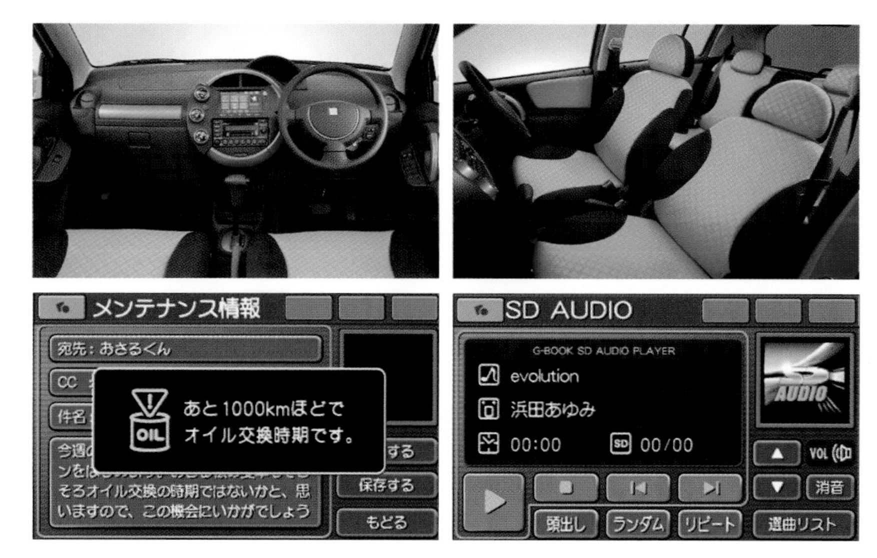
メンテナンス情報
宛先：おさるくん
CC
件名
今週の
ンを
そろオイル交換の時期ではないかと、思
いますので、この機会にいかがでしょう
あと1000kmほどで
オイル交換時期です。
保存する
もどる
SD AUDIO
G-BOOK SD AUDIO PLAYER
evolution
浜田あゆみ
00:00
00/00
AUDIO
VOL
消音
頭出し
ランダム
リピート
選曲リスト

Venturi Fétish

Design	Sacha Lakic
Engine	2.0 in-line 4
Power	134 kW (180 bhp) @ 6500 rpm
Gearbox	5-speed manual
Installation	Rear-engine/rear-wheel drive
Brakes front/rear	Discs/discs
Front tires	205/45R17
Rear tires	225/40R18
Length	3852 mm (151.7 ins)
Width	1793 mm (70.6 ins)
Height	1140 mm (44.9 ins)
Wheelbase	2544 mm (100.2 ins)
Track front/rear	1502/1505 mm (59.1/59.3 ins)
Curb weight	850 kg (1874 lb)
0–100 km/h (62 mph)	6 sec.
Top speed	225 km/h (140 mph)
Fuel consumption	8.3 ltr/100 km (28.3 mpg)

On 4 July 2001, Gildo Pallanca Pastor took control of French sports-car maker Venturi, one of the few remaining independent car manufacturers. The company, born in 1984 as MVS, went into liquidation in 2000. Venturi's new car, the provocatively named Fétish, is 100% new, an original "Grand Tourisme Light" with a deliberate emphasis on design.

The Venturi Fétish's shape looks fresh because it comes from a young and talented designer, Sacha Lakic, who works in several spheres including fashion, furniture, and even motorcycles. The Fétish is a GT in its lines and performance, but also a light car that doesn't need an especially powerful mechanical aggregate.

The design exudes femininity and sensuality, resolutely modern, yet drawing its inspiration from the past. Its outline, with its union of angles and curves, offers an alternative to mass-produced cars. This beautiful object would no doubt appeal to anyone, male or female, who wished to stand out from the crowd and assert their individuality. Inside, Plexiglas and aluminum are complemented by the Neoprene of the asymmetrical, ergonomic seats. The Fétish also benefits from a pocket-size, on-board computer that offers both GPS (global-positioning system) navigation and an MP3 reader.

The Fétish is designed to stand the test of time, unlike mass-produced cars which are constantly being superseded by newer models. It is more than a concept car; it is a real one, because orders are now being taken, at $32,880. As a counterpart to the Lotus Elise with gorgeous French design, it would be wonderful to own the Fétish, which is destined to become a collector's item.

001 FTH 75

Volkswagen Magellan

Design	Volkswagen California studio
Engine	W8
Power	202 kW (275 bhp) @ 6000 rpm
Torque	370 Nm (273 lb ft) @ 2750 rpm
Gearbox	6-speed Tiptronic
Installation	Front-engine/4-wheel drive
Front suspension	Pneumatic suspension, with active hydraulic shock absorption
Rear suspension	Pneumatic suspension, with active hydraulic shock absorption
Brakes front/rear	Discs/discs
Front tires	19 ins PAX run-flat system
Rear tires	19 ins PAX run-flat system
Length	4685 mm (184.4 ins)
Width	1860 mm (73.2 ins)
Height	1620 mm (63.8 ins)

The sales shift toward 4×4 vehicles in the last ten years, particularly in America, has hitherto escaped Volkswagen. The Magellan concept is an attempt to redress this balance.

Later this year, Volkswagen will launch its first production 4×4, the Touareg, an upmarket, five-seat vehicle aimed at Lincoln, Cadillac, Land Rover, and Lexus counterparts. The Magellan adds a third row of seats, positioning this Volkswagen against more family-oriented vehicles.

Volkswagen claims that its concept, named after the famous Portuguese explorer, is a mix of 4×4, people-carrier, and conventional wagon. As a result, the Magellan has characteristics that include a high seating position, off-road capability, and spaciousness in both the passenger area and the luggage compartment, plus the promise of agile handling.

Its front-end graphic is dominated by a newly designed grille, with the Volkswagen emblem set in the hood. This is the second Volkswagen group concept to show a radically different grille design: in 2001, Audi revealed a car with an inset grille, gaping open in the style of a pre-war racing car.

At the back, the Magellan's hatchback door and lights are smoothly integrated into its overall shape, and form trapezoidal basic elements, separated by diagonal surfaces.

Alcantara (simulated suede), leather, and aluminum dominate the cabin, which is colored in a mix of greens—light "Green Tender" and dark "Trinidad." The instruments and controls are not part of a classical cockpit, but are suspended instead in a horizontal information and control bar, between the base of the dash and the hood.

The Magellan departs from conventional Volkswagen design, and adopts a style more reminiscent of the Microbus concept. It's a fine glimpse into the future of Volkswagen.

NEVADA
535·J2P
NEVADA
535·J2P

Volkswagen Phaeton

Design	Hartmut Warkuss
Engine	6.0 W12 (3.2 V6 and V10 twin-turbo-diesel also offered)
Power	313 kW (420 bhp) @ 6000 rpm
Torque	500 Nm (369 lb ft) @ 3000 rpm
Gearbox	5-speed Tiptronic
Installation	Front-engine/4-wheel drive
Front suspension	4-link independent
Rear suspension	Trapezoidal wishbone
Brakes front/rear	Discs/discs
Front tires	235/50R18
Rear tires	235/50R18
Length	5060 mm (199.2 ins)
Width	1900 mm (74.8 ins)
Height	1450 mm (57.1 ins)
Wheelbase	2881 mm (113.4 ins)
Top speed	250 km/h (150 mph)

Volkswagen's new upper-range sedan, the Phaeton, is a car designed to lift the brand in Europe, and, with its many technical features, provide a robust challenge to the BMW 7 Series in the US.

The new Phaeton's styling is absolutely independent of other Volkswagen models. Harmonious, powerful body lines emphasize the Phaeton's imposing profile, with chrome strips providing a de-luxe flourish. Inside, except for the emblem on the four-spoke steering wheel, no detail recalls other Volkswagens.

At the top of the center console are controls for the new, four-zone Climatronic system, which permits rear-seat passenger to regulate temperatures for the left- and right-hand back seats, in addition to allowing individual control for front-seat occupants. The Climatronic system also funnels draft-free cooling or heating air into the interior via automatically opening and closing vents. Integral humidity control automatically prevents fogged-up windows.

The Phaeton comes with two high-performance, gasoline engines. Top of the line is the new W12, an incredible, 6 ltr, twelve-cylinder engine with abundant power and torque. There is also the new V10 TDI, which ensures that the Phaeton is equipped with the world's most powerful, and tractable, passenger-car diesel engine.

Volkswagen is now seriously challenging BMW and Mercedes-Benz with its Phaeton, even slightly overshadowing its own Audi luxury brand with the new car. Most of all, it is pushing out the public perception of what constitutes a Volkswagen. If VW discovers that a truly premium product can be sold without a premium brand name, it is opening up a world of new market possibilities.

WOB JA 500

WOB JA 500

Volkswagen Polo

Design	Hartmut Warkuss
Engine	1.4 in-line 4 (1.2 and 1.4 in-line 4, and 1.9 in-line 4 diesel, also offered)
Power	55 kW (75 bhp) @ 5000 rpm
Torque	126 Nm (93 lb ft) @ 3800 rpm
Gearbox	5-speed manual/4-speed automatic
Installation	Front-engine/front-wheel drive
Front suspension	MacPherson strut
Rear suspension	Trailing arm with torsion beam
Brakes front/rear	Discs/discs, ABS
Front tires	165/70R14
Rear tires	165/70R14
Length	3897 mm (153.4 ins)
Width	1650 mm (65 ins)
Height	1465 mm (57.7 ins)
Wheelbase	2460 mm (96.9 ins)
Track front/rear	1435/1425 mm (56.5/56.1 ins)
Curb weight	1183 kg (2609 lb)
0–100 km/h (62 mph)	12.9 sec.
Top speed	172 km/h (107 mph)
Fuel consumption	6.5 ltr/100 km (36.1 mpg)
CO_2 emissions	156 g/km

In its fourth generation, the Volkswagen Polo has grown up. It is now 153 mm (6 ins) longer than the previous model, and its wheelbase has been extended by 51 mm (2 ins). With the small Lupo beneath it in the VW-model pecking order, Volkswagen can use its Polo to target customers wanting either a compact family car that is cheaper than a Golf, or a high-quality, urban runabout. The Polo also follows a traditional pattern of established models expanding as they mature, making room for sub-sector cars below them.

Polo design has never been about being radical, so the new model remains reassuringly and sensibly Teutonic. Gentle, body-colored surfaces are mixed with occasional black features, including the lower trim, the glass blackouts, and the front grille. The headlamps are now twin ellipses, like the Lupo's, while the rear is obviously Volkswagen Golf. The tailgate wraps around the body, and the lamps are now mostly red, with a central, high-mounted brake lamp (the car-design argot for which is now CHMSL) in the roof spoiler.

The new Polo's technological features include an Electronic Stability Program (ESP), incorporating Hydraulic Brake Assist (HBA), which cuts emergency-braking distance. "Climatic," a fuel-saving semi-automatic air-conditioning system, and integrated rear child seats are among the options offered.

The new Polo is a quality small car packed with safety features. Volkswagen has been careful to update its aesthetics only slightly, bringing it into line with such other Volkswagen hatchbacks as the Lupo and the Golf. This high-selling model is in the enviable position of being an almost guaranteed success, so its designers were anxious not to take risks.

POLO
WOB DJ 120

Volkswagen W12 Coupé

Design	Italdesign
Engine	6.0 W12
Power	447 kW (600 bhp) @ 7000 rpm
Torque	620 Nm (457 lb ft) @ 5800 rpm
Gearbox	Sequential 6-speed manual
Installation	Mid-engine/4-wheel drive
Front suspension	Double wishbone
Rear suspension	Double wishbone
Brakes front/rear	Discs/discs, ESP
Front tires	255/35ZR19
Rear tires	275/40ZR19
Length	4550 mm (179.1 ins)
Width	1920 mm (75.6 ins)
Height	1100 mm (43.3 ins)
Wheelbase	2630 mm (103.5 ins)
Curb weight	1200 kg (2646 lb)
0–100 km/h (62 mph)	3.5 sec.
Top speed	349 km/h (217 mph)

Dr Ferdinand Piech, the recently retired chairman of Volkswagen, liked to push the envelopes of the brands he controlled within the group, with Volkswagen itself now covering everything from a three-cylinder economy car to a luxurious executive limousine. Even so, a mid-engine, W12-powered, two-seat supercar seemed an extremely unlikely addition to the Wolfsburg range until now, that is, and the revelation of the virtually customer-ready W12 Coupé at the 2001 Tokyo Motor Show.

The W12 Coupé, with a body design from Italdesign, has been refined from previous concept models, the main difference being new front and rear lamps. It's an out-and-out sports car that differs from, say, a Ferrari in not having such extrovert styling: Volkswagen's restrained body-surface language has been cleverly retained by Italdesign.

The W12 has two forward-tilting doors, and a central glass panel extending the length of the roof, from the windshield to the very end of the engine compartment, so the magnificent W12 power unit is clearly displayed through this transparent engine cover. At the heart of the Coupé is the W12 engine (a production first) that can develop 447 kW (600 bhp), making it one of the fastest sports cars. The very necessary rear spoiler automatically activates above 120 km/h (75 mph) for additional downforce at such high speeds.

A luxury, lightweight, high-tech, but sparing look for the interior is achieved by using leather, aluminum, and carbon fiber. The instrument layout is classic sports car, with two big, round, main displays providing the most important information. Controls for air conditioning, navigation system, on-board computer, and car telephone are controlled via a color-coded display in the center console.

The W12 Coupé has already proved that it is more than just an exciting-looking machine: in October 2001 it set the world speed record for distance covered in twenty-four hours. On the Nardo high-speed circuit in southern Italy, the car covered 7088.5 km (4402 miles) at an average speed of 295.4 km/h (183 mph), bettering the previous record set by a Chevrolet Corvette LTS by 12.1 km/h (7.5 mph). Proof, if any were needed, that Volkswagen is as serious about making the W12 a world-class product as it is about its Golf and Passat.

Volvo XC90

Design	Peter Horbury
Engine	2.9 in-line 6 (2.5 in-line 5 and 2.5 in-line 5 turbo-diesel also offered)
Power	200 kW (268 bhp) @ 5200 rpm
Torque	380 Nm (280 lb ft) @ 1800 rpm
Gearbox	4-speed automatic, with clutchless manual facility
Installation	Front-engine/4-wheel drive
Front suspension	MacPherson strut
Rear suspension	Multi-link independent
Brakes front/rear	Discs/discs
Front tires	235/65R17
Rear tires	235/65R17
Length	4800 mm (189 ins)
Width	1890 mm (74.4 ins)
Height	1740 mm (68.5 ins)
Wheelbase	2859 mm (112.6 ins)
Track front/rear	1634/1624 mm (64.3/63.9 ins)
Curb weight	2110 kg (4653 lb)

Volvo has largely missed out on the 4×4 craze that has swept the world's car markets, but the XC90 will rectify that.

Larger than the V70 wagon, in side profile it shares a similar outline to that of the BMW X5 4×4. The Volvo, however, brings its own distinctive lines—a styling direction created by chief designer Peter Horbury—to the 4×4 shape.

The muscular stance features such key Volvo design cues as a bold and upright nose with egg-crate grille, stepped hood, and pronounced shoulder to define the body line. Highly detailed, bold, and complex-shaped headlamps are a departure for Volvo, although the tail lamps are more characteristic, extending up the rear pillars to improve visibility to other drivers. Other detailing, such as dark plastic bumpers and sill guards, emphasizes the 4×4 look.

The cleverest feature of the XC90 is its interior package—the choice of body architecture, and how the interior space is used. The passenger cabin is moved as far forward as possible. This is helped by Volvo's use of space-saving, transverse-mounted engines, while competitors such as BMW and Mercedes-Benz employ less space-efficient, longitudinal engines. As a result, Volvo's designers have squeezed a third row of seats into a vehicle just 4.8 m (15.7 ft) long. As might be expected from a company with a reputation for safety, all seats are protected by airbags.

Scandinavian design simplicity is the theme of the interior—as it is in Volvo's sedans and wagons—with simple lines, rolling surfaces, and gentle color contrasts. A vast array of customizing accessories is available, including aluminum running boards, a color-coded body kit, a rear skid plate, and roof bars to carry lifestyle equipment.

VOLVO
SRM 177
XC90

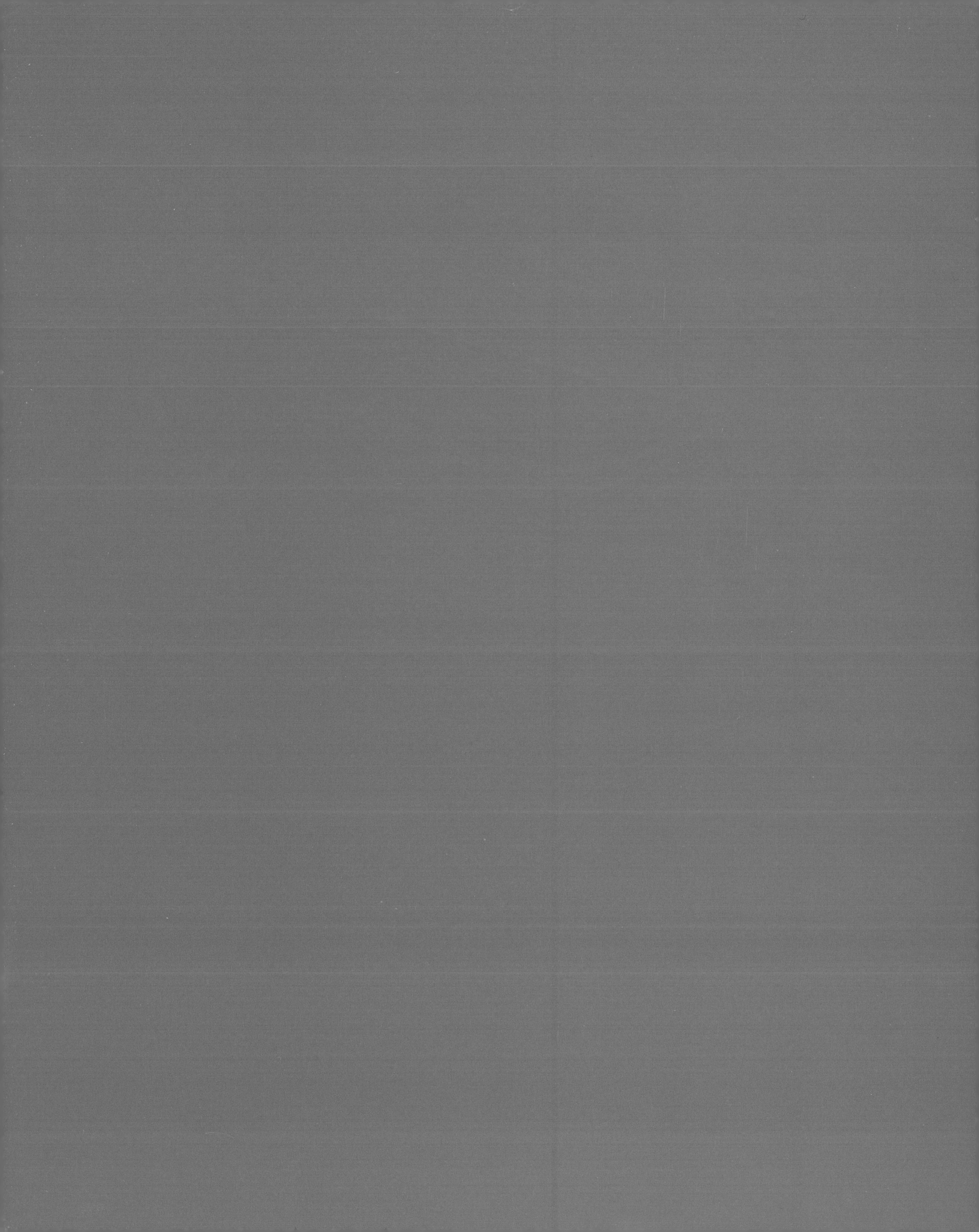

Profiles of Key Designers

Ian Callum

Ian Callum and his brother Moray are not only Scotland's best-known car designers, but also among the very few siblings involved in the international car-styling business. Born in Dumfries, as youngsters they were inspired by reading car magazines. Ian, born in 1954, loved the Jaguar XJ6, and still recalls its debut vividly. His all-time favorite car is the Ferrari 250GT SWB—which he still sometimes draws for "relaxation."

Ian won a place on a transport-design course at Lanchester Polytechnic, and then studied industrial design at Glasgow School of Art:

"It was wonderful. I had great support from the college, and I had contact with car companies. I think people should have a better understanding of design generally, rather than try to go straight into designing cars."

After gaining a master's degree in Industrial Design at London's Royal College of Art in 1978, Ian spent the next twelve years globe-trotting around Ford design centers in Britain, Japan, the US, Australia, and Germany, before being appointed design manager responsible for the Ford-owned Ghia design studio in Turin. There he helped design the Ford Fiesta, Mondeo, RS200, Escort RS Cosworth, and Probe, as well as concept cars including the Via, and the Zig and Zag two-seat compacts.

On returning to Britain in 1990, Ian, now married with two children, became general manager and chief designer at TWR, where he worked on projects for Aston Martin, Holden, Nissan, Mazda, Range Rover, and Volvo. He was responsible for the designs of a selection of sports cars, including the Nissan R390, Aston Martin DB7, Aston Martin Vanquish, and Volvo C70. The DB7, star of the 1993 Geneva International Motor Show, was named one of the most beautiful cars in the world by a panel of Italian artists, architects, and photographers.

In August 1999, Jaguar announced Ian Callum's appointment as director of styling, where he has total responsibility for shaping the brand.

The first of his new designs was the R-Coupé concept, revealed at the 2001 Frankfurt Motor Show. Ian Callum says: "The exterior of the Jaguar R-Coupé looks engineered, not merely styled. Its integrity comes from strong, geometric forms such as the tubular fuselage—recalling that of the Jaguar E-type."

above from top
Ian Callum
Jaguar R-Coupé, 2002
Nissan R390 GTi, 1998

opposite, clockwise from top
Aston Martin Vanquish, 2001
Aston Martin DB7, 1993
Volvo C70, 1996

A002 UMV

V12 AML

Giorgetto Giugiaro

Giorgetto Giugiaro is a living legend in car design, with a career stretching back to the 1950s, and a portfolio that has ridden the crest of contemporary car design ever since. He is the father of the modern family car, with all its practicality and everyday style.

He was born in 1938 in Garessio, north-west Italy. Both his father and grandfather had been painters, and at the age of seventeen Giorgetto Giugiaro began work in the Fiat styling studio as a prolific producer of renderings—some fanciful, some mundane—for future Fiat models.

In 1959, aged just twenty-one, he left to become head of design at Bertone. Life at the venerable coachbuilding company was hectic, and young Giugiaro's work flowed in the shape of countless show cars, and several successful production models. The Fiat 850 Spider was probably the best known internationally, after American enthusiasts took it to their hearts.

In 1965 Giugiaro left Bertone to join Ghia, and four years later he established his own design consultancy, ItalStyling, which later became Italdesign. He never looked back.

Giugiaro has styled over one hundred cars, many of which have been best-selling, mass-market models. The 1971 Alfa Romeo Alfasud, 1974 Volkswagen Golf, 1975 Hyundai Pony, 1979 Lancia Delta, and 1980 Fiat Panda are now seen as definitive interpretations of the practical family hatchback, the last four of them displaying sharp, "folded-paper" lines. He applied the same touch to sports cars, such as the 1972 Lotus Esprit, and the 1978 De Lorean DMC-12.

Giugiaro's intuition for the demands of contemporary design continues unabated. In recent years he has produced the Fiat Punto, Seat Ibiza, and—to acclaim on both sides of the Atlantic—the Lexus GS300. He has won a number of prestigious design awards. In 1984, he was awarded an honorary degree from the Royal College of Art, London. In 1999, in Las Vegas, a jury of more than 120 journalists elected him Designer of the Century. And in 2001, together with twelve other automotive greats, he was immortalized in the "hall of fame" at the Palexpo exhibition in Geneva.

Giorgetto Giugiaro is one of the greatest car designers of the twentieth and—now—the twenty-first centuries. With such stunning models as the recently launched Brera, his designs show no signs of becoming stale.

above from top
Giorgetto Giugiaro
Fiat Panda, 1980

opposite, clockwise from top left
Italdesign Brera, 2002
Fiat Punto, 1993
Volkswagen Golf, 1974
Volkswagen Golf, 1974
Lotus Esprit, 1974

DVV 712T

Freeman Thomas

Car design that gets non-fanatics talking is what every modern manufacturer dreams of, but something that very few achieve. Thanks to 45-year-old Freeman Thomas, however, the images of Audi, Volkswagen, and soon, in all probability, Chrysler have been raised. Even non-drivers point and stare at the Audi TT, and the rejuvenated Volkswagen Beetle, cars that soften even the most ardent anti-car campaigners just a little.

This puller of heartstrings is the new head of Chrysler's Pacifica Center in Carlsbad, California. In recognition of his visionary genius, he is also vice-president of Chrysler's product-design strategy. Thomas is responsible for creating a dedicated advanced bureau to foster new products, to plan strategy, and to oversee all future design at Chrysler.

Thomas started his car-styling career at Porsche in 1983, after graduating in industrial design from the Art Center College of Design, California. He worked initially as a senior member of the Porsche 959 and 965 project teams, before departing in 1987 to work as an independent consultant. In 1991, he joined Audi as chief designer, and, while there, penned the Audi TTS Spyder Concept, unveiled at the 1995 Tokyo Motor Show, and the Audi TT Concept for the 1995 Frankfurt Motor Show. Thomas can claim joint responsibility, with J. Mays, for designing the fabulous new Beetle, in 1994, in his position as chief designer at the Volkswagen–Audi design center.

He joined DaimlerChrysler in June 1999, with a mission to revive American design to the heights achieved during the company's 1950s heydays. There is useful synergy between Chrysler's established plundering of its heritage and Thomas's sharp-eyed regard for repackaged nostalgia.

Chrysler has recently shown some bold new designs, including the PT Cruiser convertible, and the Pacifica concept. This is an excellent base on which Freeman Thomas can build some exciting new Chrysler models. We should expect a few surprises in 2003–04.

above, clockwise from top right
Freeman Thomas
Chrysler PT Cruiser, 1999
Chrysler Pacifica, 2002
Chrysler Pacifica, 2002

opposite, clockwise from top left
Audi TT, 1995
Audi TT Coupé, 1995
Volkswagen new Beetle, 1997

Audi

Technical Glossary

The following list explains the terminology used in the specification tables that accompany the model descriptions. The amount of data available for any given model depends on its status as a concept or a production car. More information is usually available for models currently in or nearing production.

Design	Where possible, the name of the designer or the design studio responsible for the car is given.
Engine	Engine size is quoted in liters, and refers to the swept volume or displacement of the cylinders per crankshaft rotation; 6.0, for example, means a 6 ltr engine—also known as a 6000cc engine, standing for cubic centimeters. "In-line" or "V" followed by a number refers to the engine's number of cylinders. An in-line 4 engine has four cylinders in a single row, while a V8 engine has eight cylinders arranged in a V-formation. In-line engines of more than six cylinders are rare today because they take up too much packaging space—an in-line 12, for instance, would require a very long hood. Only Volkswagen makes a W12, an engine with its 12 cylinders arranged in a W-formation. The cylinder configuration is usually chosen on cost grounds: the higher the car's retail price, the more cylinders product planners can include.
Power	Engine power is given both in kilowatts (kW) and brake horsepower (bhp)—the current and recent measurements. Both are calculated at optimum engine crankshaft speed, given in "rpm" (revolutions per minute) by manufacturers as a net measurement—in other words, an engine's output after power has been absorbed by other equipment and the exhaust system—and measured by a "brake" applied to the drive shaft.
Torque	In car terms, "torque" means pulling power, generated by twisting force from the engine's crankshaft. It is given in Newton-meters (Nm) and pounds-feet (lb ft). The better the torque, the more force the engine can apply to the driven wheels.
Gearbox	This is the mechanical means by which power is transmitted from the engine to the driven wheels. There is a wide variety of manual (with a clutch) and automatic (clutchless) versions. A recent trend has been for clutchless manual systems, called "automated manual," and automatics with the option to change gear manually, sometimes branded "Tiptronic," "Steptronic," or "Easytronic." A CVT (continuously variable transmission) is an automatic with a single "speed": the system uses rubber or steel belts to take engine power to the driven wheels, with drive pulleys that expand and contract to vary the gearing. A "single-speed gear-reduction drive," as fitted to the Th!nk City, is a special transmission for an electric motor. A "sequential manual" is a manual gearbox with pre-set gear ratios that are selected sequentially.
Suspension	All suspension systems cushion the car against road or terrain conditions to maximize comfort, safety, and roadholding. Heavy and/or off-road vehicles use "rigid axles" at the rear or front/rear. These are suspended, using robust, leaf-type springs, and steel "wishbones," or "A-arms," with "trailing arms." "Watts linkages" and "Panhard rods" are suspension components that help keep cars stable or improve handling. "Semi-rigid axles" are often found at the back of front-wheel-drive cars, in conjunction with a "torsion-beam" trailing-arm axle. "Independent" suspension means that each wheel can move up and down on its own, often with the help of "trailing arms" or "semi-trailing arms." A "MacPherson strut," named after its inventor, Earl MacPherson, a Ford engineer is a suspension upright, fixed to the car's structure above the top of the tire. It carries the wheel hub at the bottom, and incorporates a hydraulic damper. It activates a coil spring, and, when fitted at the front, turns with the wheel.

Brakes	Almost all modern cars feature disc brakes for all four wheels. A few, low-powered models still feature drum brakes at the back for cost reasons. "ABS" (anti-lock braking system) is increasingly fitted to all cars: it regulates brake application to prevent the brakes from locking up in an emergency or slippery conditions. "Brake Assist" is a system that does this electro-hydraulically. EBD (electronic brake-force distribution) is a pressure regulator that, in braking, spreads the car's weight more evenly, so locked-up brakes are avoided. "ESP" (electronic stability program) is Mercedes-Benz's electronically controlled system that helps keep the car pointing in the right direction at high speeds; sensors detect wayward roadholding, and apply the brakes indirectly to correct it. "Dynamic stability" is a similar system. "Brake-by-wire" is a totally electronic braking system that sends signals from brake pedal to brakes with no intermediate hydraulic or mechanical actuation. "TCS" (traction-control system) is a feature that holds acceleration slip within acceptable levels to prevent wheelspin, and, therefore, improves adhesion to the road.
Tires	The size and type of wheels and tires are given in the internationally accepted formula. Representative examples include: 315/70R17, 235/50VR18, 225/50WR17, 235/40Z18, and 225/40ZR18. In all cases, the first number before the slash is the tire width in millimeters. The number after the slash is the height-to-width ratio of the tire section as a percentage. The letter R denotes radial construction (cross-ply tires are still produced, but not for cars). Letters preceding R are a guide to the tire's speed rating, denoting the maximum safe operating speed. H tires can be used at speeds up to 210 km/h (130 mph), V up to 240km/h (150 mph), W up to 270 km/h (170 mph), and Y up to 300 km/h (185 mph). The last number is the diameter of the wheel in inches. A PAX is a wheel-and-tire in one unit, developed by Michelin, and fitted to, for example, the Volkswagen Magellan (19/245 PAX means a 19 ins wheel with a 245 mm tire width). The rubber tire element is clamped to the steel wheel part, rather than held on by pressure. The height of the tire walls is reduced, which can free up space for better internal packaging, or allow bigger wheels for sexier, concept-car looks. A PAX can also run flat for 200 km (125 miles), at 80 km/h (50 mph), thus eliminating the need for a spare.
Wheelbase	The distance between the center of the front wheel and center of the rear wheel.
Track front/rear	The distance between the center of the front or rear tires across the car, measured at the ground.
Curb weight	The amount a car weighs with a tank of fuel, all oils, and coolants filled, and all standard equipment, but no occupants.
CO_2 emissions	Carbon dioxide emissions, which are a direct result of fuel consumption. Less than 100 g/km is a very low emission, 150 g/km is good, 300 g/km is bad. CO_2 contributes to the atmospheric "greenhouse effect."

Other car design terms explained

A-, B-, C-, D-pillars	Vertical roof-support posts that form parts of a car's bodywork. Moving rearward, the A-pillar sits between windshield and front door, the B-pillar between front and rear doors, the C-pillar between rear doors and rear window, hatchback or wagon rear side windows, and the D-pillar (on a wagon) between rear side windows and tailgate. Confusingly, however, some designs refer to the central pillar between front and rear doors as a B-pillar where it faces the front door, and a C-pillar where it faces the rear one.

Cantrail	The structural beam that runs along the tops of the doors.
Coefficient of drag	Also known as Cd, this is the result of the complex scientific equation that proves how aerodynamic a car is. The Audi A2 has a Cd of 0.28, but the Citroën SM of thirty years ago measured just 0.24, so it is not exactly rocket science. Drag means the resistance of a body to airflow. Low drag means better penetration, less friction with the atmosphere, and therefore more efficiency, although sometimes poor dynamic stability.
Drive-by-wire technology	Increasingly featured on new cars, these systems do away with mechanical elements, and replace them by wires transmitting electronic signals to activate functions such as throttle, brakes, and steering.
Drivetrain	The assembly of "organs" that gives a car motive power: engine, gearbox, driveshaft, wheels, brakes, suspension, and steering. This is also loosely known as a chassis, and can be transplanted into several different models to save on development costs.
Feature line	A styling detail usually added to a design, such as the Cadillac Cien, deliberately to differentiate it from its rivals, and generally not related to functional areas such as, for instance, door apertures.
Greenhouse	Car-design slang for the glazed area of the passenger compartment that usually sits above the car's waist level.
Instrument panel	The trim panel that sits in front of the driver and front passenger, also known as "dashboard," or "dash."
Monospace/"one-box"	A "box" is one of the major volumetric components of a car's architecture. In a traditional sedan, there are three boxes: one for the engine, one for the passengers, and one for the luggage. A hatchback, missing a trunk, is a two-box car, while a large MPV, such as the Citroën C8, is a one-box design, or "monospace."
MPV	Short for multi-purpose vehicle, applied to tall, spacious cars that can carry at least five passengers, and often as many as nine, or versatile combinations of people and cargo. The 1983 Chrysler Voyager and 1984 Renault Espace were the first. The 1977 Matra Rancho was the very first "mini-MPV," and the 1991 Mitsubishi Space Runner was the first in the modern idiom.
Platform	The invisible, but elemental, and expensive, basic structure of a modern car. It is the task of car designers to achieve maximum aesthetic diversity from a single platform.
Powertrain	The engine, gearbox, and transmission "package" of a car.
Spaceframe	A structural frame that supports a car's mechanical systems and cosmetic panels.
SUV	Short for sports utility vehicle, this is a four-wheel-drive car designed for leisure off-road driving, but not necessarily agricultural or industrial use. Therefore, a Land Rover Defender is not an SUV, while a Land Rover Freelander is. The line between the two is sometimes difficult to draw, and identifying a pioneer is tricky: SUVs as we know them today were defined by Jeep in 1986 with the Wrangler, Suzuki in 1988 with the Vitara, and Daihatsu in 1989 with the Sportrak.
Targa	Porsche had been very successful in the Targa Florio road races in Sicily, so, in celebration, in 1965 the company applied the name "Targa" (Italian for "shield") to a new 911 model that featured a novel, detachable roof panel. It is now standard terminology for the system, although a Porsche-registered trademark.
Telematics	Any individual communication to a car from an outside base station, for example, satellite navigation signals, automatic emergency calls, roadside assistance, traffic information, and dynamic route guidance.

New York International Auto Show
13–22 April 2001

Concept
Hummer H2 SUT
Lincoln MK 9
Suzuki SX

Frankfurt Motor Show
11–23 September 2001

Concept
Audi Avantissimo
Citroën C-Crosser
Ford Fusion
Hyundai Clix
Jaguar R-Coupé
MCC Smart Tridion4
MG X80
Nissan Crossbow
Nissan mm.e
Opel Frogster
Opel Signum2
Pininfarina Ford Start
Renault Talisman
Saab 9X
Seat Tango
Toyota ES³

Production
BMW 7 Series
Cadillac CTS
Citroën C3
Daewoo Kalos
Ford Fiesta
Hyundai Tiburon Coupé
Lamborghini Murciélago
Maserati Spyder
Mercedes-Benz SL
Mercedes-Benz Vaneo
Skoda Superb
Toyota Corolla
Volkswagen Polo

Tokyo Motor Show
24 October – 7 November 2001

Concept
Daihatsu Copen
Daihatsu FF Ultra Space
Daihatsu Muse
Daihatsu U4B
Daihatsu UFE
Honda Bulldog
Honda Dualnote
Honda Unibox
Isuzu Zen
Mazda Secret Hideout
Mercedes-Benz F400 Carving
Mitsubishi CZ2
Mitsubishi Space Liner
Mitsubishi SUP
Nissan GT-R
Nissan Ideo
Nissan Kino
Nissan Moco
Nissan Nails
Toyota DMT
Toyota FXS
Toyota Ist
Toyota pod
Toyota RSC
Toyota Voltz
Toyota WiLL VC

Production
Honda CR-V
Mazda6/Atenza
Mazda RX-8
Nissan Z
Volkswagen W12 Coupé

Greater LA Auto Show
3–13 January 2002

Concept
Isuzu GBX
Lexus Movie
Lincoln Continental

Production
Chrysler Crossfire
Hummer H2
Lincoln Navigator
Pontiac Vibe
Th!nk City

North American International Auto Show (NAIAS)
Detroit
6–21 January 2002

Concept
Acura RD-X
Cadillac Cien
Chevrolet Bel Air
Chrysler Pacifica
Dodge M80
Dodge Razor
Ford F-350 Tonka
Ford GT40
GM AUTOnomy
Infiniti FX45
Jeep Compass
Mercedes-Benz Vision GST
Mitsubishi SUP Cabrio
Nissan Quest
Pontiac Solstice
Saab 9-3X
Toyota ccX
Volkswagen Magellan

Production
Cadillac XLR
Chevrolet SSR
Honda Pilot
Infiniti G35
Lexus GX470
Range Rover
Subaru Baja
Suzuki Aerio
Volvo XC90

Brussels International Motor Show
17–27 January 2002

Production
Mercedes-Benz E-Class

Chicago Auto Show
8–17 February 2002

Production
Kia Sorento

72nd Geneva International Motor Show
5–17 March 2002

Concept
Bertone Novanta
BMW CS1
Fioravanti Yak
Irmscher Inspiro
Italdesign Brera
Maybach
Mazda MX Sport Runabout
Mitsubishi Pajero Evolution 2+2
Nissan Yanya
Opel Concept M
Peugeot RC
Renault Espace
Rinspeed Presto
Rover TCV
Skoda Tudor
Toyota UUV
Venturi Fétish

Production
Citroën C8
Fiat Ulysse
Hyundai Getz
Koenigsegg CC 8S
Lancia Phedra
Maserati Coupé
Matra m72
MCC Smart Crossblade
Mercedes-Benz CLK
Opel/Vauxhall Vectra
Peugeot 807
Volkswagen Phaeton

Major International Auto Shows 2002–2003

Paris Motor Show (Mondial de l'automobile)
Saturday 13 September – Sunday 13 October 2002
Paris Expo, Paris, France
www.mondialauto.tm.fr

Budapest Motor Show
Wednesday 2 – Sunday 6 October 2002
Budapest Fair Center, Budapest, Hungary
www.automobil.hungexpo.hu

Prague Auto Show (Autoshow Praha)
Thursday 24 – Sunday 27 October 2002
Prague Exhibition Grounds, Prague, Czechoslovakia
www.incheba.cz

British International Motor Show
Friday 25 –Thursday 31 October 2002
National Exhibition Centre (NEC), Birmingham, UK
www.motorshow.co.uk

Seoul Motor Show
Thursday 21 – Friday 29 November 2002
Convention and Exhibition Center (COEX), Seoul, Korea
www.motorshow.or.kr

Greater LA Auto Show
Saturday 4 – Sunday 12 January 2003
Los Angeles Convention Center, Los Angeles, USA
www.laautoshow.com

North American International Auto Show (NAIAS)
Saturday 11 – Monday 20 January 2003
Cobo Center, Detroit, USA
www.naias.com

Brussels International Motor Show
Saturday 18 – Sunday 26 January 2003
Brussels Expo, Brussels, Belgium
www.febiac.be/motorshows

Amsterdam Motor Show (AutoRAI 2003)
Friday 7 – Sunday 16 February 2003
Amsterdam RAI Convention Center, Amsterdam, The Netherlands
ww.autorai.nl

Brisbane Motor Show
Friday 7 – Sunday 16 February 2003
Brisbane Convention and Exhibition Centre, South Bank, Brisbane, Australia
www.brisbanemotorshow.com.au

Canadian International Auto Show
Friday 14 – Sunday 23 February 2003
SkyDome, Toronto, Canada
www.autoshow.ca

Chicago Auto Show
Friday 14 – Sunday 23 February 2003
McCormick Plaza South & Metro Toronto Convention Center, Chicago, USA
www.chicagoautoshow.com

Melbourne International Motor Show
Thursday 27 February – Monday 10 March 2003
Melbourne Exhibition Centre, Melbourne, Australia
www.motorshow.com.au

73rd Geneva International Motor Show
Thursday 6 – Sunday 16 March 2003
Palexpo, Geneva, Switzerland
www.salon-auto.ch

New York International Auto Show
Friday 18 – Sunday 27 April 2003
The Jacob Javits Convention Center, New York, USA
www.autoshowny.com

Tallinn Motor Show (Motorex 2003)
Wednesday 23 – Sunday 27 April 2003
The Estonian Fairs Center
Tallinn, Estonia
www.fair.ee

International Motor Show, Barcelona
Saturday 26 April – Sunday 4 May 2003
Fira de Barcelona, Barcelona, Spain
www.salonautomovil.com

The New London Motor Show
Including the International Car Designer of the Year Award
Saturday 21 – Sunday 29 June 2003
Earls Court Exhibition Centre, London, UK
www.londonmotorshow.co.uk

Frankfurt Motor Show
Saturday 13 – Sunday 21 September 2003
Messe Frankfurt GmbH, Frankfurt am Main, Germany
www.iaa.de

Paris Motor Show (Mondial de l'automobile)
Sunday 28 September – Monday 13 October 2003
Paris Expo, Paris, France
www.mondialauto.tm.fr

Middle East International Motor Show
Monday 6 – Friday 10 October 2003
Dubai World Trade Center (DWTC), Dubai, United Arab Emirates
www.dubaimotorshow.com

Tokyo Motor Show
Saturday 25 October – Wednesday 5 November 2003
Nippon Convention Center, Chiba City, Tokyo
www.motorshow.or.jp

Riyadh Motor Show
Sunday 9 – Friday 14 November 2003
Riyadh Exhibition Center, Riyadh, Saudi Arabia
www.recexpo.com

Marques and their Parent Companies

The automotive industry is one of the world's biggest and is evolving constantly. Its centres have traditionally been North America, northern Europe, and Japan, with outposts in South America, Australia and South Korea. Growth areas for both customers and manufacturing enterprise, however, are seen as India, China, Russia and former Eastern Bloc territories. From hundreds of separate car-making companies, the industry has consolidated into ten groups: General Motors, Ford, DaimlerChrysler, VW, Toyota, Peugeot, Renault, BMW, Honda and Hyundai account for at least nine of every ten cars produced globally today. In the late 1990s a corporate feeding frenzy saw smaller companies and marques rapidly absorbed by these industrial giants: the remaining independents tend to be so because their lack of scale economies point to an uncertain future. The homogenization of the car industry has, however, put a renewed emphasis on design: an ever-dwindling number of basic car types means that manufacturers rely heavily on design differentiation for customer recognition.

BMW
BMW
Mini
Riley*
Rolls-Royce (from 2003)
Triumph*

DaimlerChrysler
Chrysler
De Soto*
Dodge
Hudson*
Imperial*
Jeep
Maybach
Mercedes-Benz
Mitsubishi
Nash*
Plymouth*
Smart

Fiat Auto
Abarth*
Alfa Romeo
Autobianchi*
Ferrari
Fiat
Innocenti*
Lancia
Maserati

Ford
Aston Martin
Daimler
Ford
Jaguar
Land Rover
Lincoln
Mazda
Mercury
Range Rover
Th!nk
Volvo

General Motors
Buick
Cadillac
Chevrolet
Daewoo
GMC
Holden
Hummer
Isuzu
Oldsmobile*
Opel
Pontiac
Saab
Saturn
Subaru
Vauxhall

Honda
Acura
Honda

Hyundai
Asia Motors
Hyundai
Kia

MG Rover
Austin*
Austin-Healey*
MG
Morris*
Rover
Wolseley*

Peugeot
Citroën
Hillman*
Humber*
Panhard*
Peugeot
Simca*
Singer*
Sunbeam*
Talbot*

Proton
Lotus
Proton

Renault
Alpine*
Dacia
Datsun*
Infiniti
Nissan
Renault
Renault Sport

Toyota
Daihatsu
Lexus
Toyota

VW
Audi
Auto Union*
Bentley
Bugatti*
Cosworth
DKW*
Horch*
Lamborghini
NSU*
Rolls-Royce (until 2003)
Seat
Skoda
Volkswagen
Wanderer*

Independent marques
Bertone
Bristol
Caterham
Fioravanti
Heuliez
Irmscher
Italdesign
Lada
Koenigsegg
Matra
Mitsuoka
Morgan
Pininfarina
Porsche
Rinspeed
SsangYong
Subaru
Suzuki
Tata
TVR
Venturi
Westfield
Zagato

* Dormant marques

Picture Credits

The illustrations in this book have been reproduced with the kind permission of the following manufacturers:

Alfa Romeo
Audi AG
Automobili Lamborghini
Bertone SpA
BMW AG
Cadillac
Chevrolet
Citroën Communication
Daewoo Motors
Daihatsu Motor Co. Ltd
DaimlerChrysler
Fiat Auto
Fioravanti srl
Ford Motor Company
General Motors
Honda Motor Co.
Hummer
Hyundai Car UK Ltd
Irmscher Automobilbau
Isuzu
Italdesign
Jaguar Cars Ltd
Kia
Koenigsegg
Lancia
Land Rover
Lincoln
Maserati SpA
Matra
Mazda Motors
MG Rover Group Ltd
Mitsubishi Motors Corporation
Nissan Motors
Opel
Peugeot
Pininfarina
Pontiac
Renault
Rinspeed
Saab Automobile AB
Seat SA
Skoda Auto
Subaru
Suzuki Motor Corporation
Th!nk
Toyota Motor Corporation
Venturi Automobiles
Volkswagen
Volvo Car Corporation

Acknowledgments

I am particularly grateful to Merrell Publishers for having the vision to see the potential of this book in its early stages, and for their unwavering professionalism in helping to deliver it. I must especially thank Julian Honer, Kate Ward, Anthea Snow, and Emily Sanders. Thanks are also due to the manufacturers' public relations offices, and especially those based in the UK, which have been so supportive in supplying images and technical information.

Many friends both within and outside the automotive industry have given invaluable advice. In particular, I would like to thank Vicky Gallagher for her continual support and encouragement, and Peter Newbury for his persistence in researching the technical specifications. I must also thank Giles Chapman, Tony Lewin, Julian Rendell, and Karl Ludvigsen for their expert and professional editorial support.

Finally, I should like to dedicate this book to my parents, who always give unconditional support to my projects.

Stephen Newbury
Cheltenham, 2002